HELEN PLUM MEMORIAL LIBRARY
3 1502 00765 5564

P9-AZW-374

# A Piece of the Sun

# A Piece of the Sun

## THE QUEST FOR FUSION ENERGY

**DANIEL CLERY**

**OVERLOOK DUCKWORTH**
**NEW YORK • LONDON**

This edition first published in hardcover in the United States and the United Kingdom in 2013 by
Overlook Duckworth, Peter Mayer Publishers, Inc.

New York
141 Wooster Street
New York, NY 10012
www.overlookpress.com
For bulk and special sales, please contact sales@overlookny.com,
or write us at the address above

London
30 Calvin Street
London E1 6NW
info@duckworth-publishers.co.uk
www.ducknet.co.uk



Cataloging-in-Publication Data is available from the Library of Congress

*Book design and typeformatting by Bernard Schleifer*
Manufactured in the United States of America
ISBN US: 978-1-4683-0493-0
ISBN UK: 978-0-7156-4525-3

*To Bernadette*
*who made it all possible*

*And to Sam and Ellen*
*for their boundless enthusiasm*

# Contents

# A Piece of the Sun

CHAPTER 1

# Why Fusion?

We owe everything to fusion. Our own Sun and every star that shines in the night sky are powered by fusion. Without it, the Cosmos would be dark, cold and lifeless. Fusion fills the Universe with light and heat, and allows life to happen on Earth and probably elsewhere. The Earth itself, the air we breathe and the very stuff we are made of are the products of fusion.

Following the Big Bang, once things had cooled down enough for neutral atoms to form, the Universe was filled with a fairly even distribution of hydrogen, the simplest atom. There was a bit of helium and some of the mysterious dark matter, but the Cosmos appeared to be just hydrogen atoms and empty space. So how did fusion transform this blank canvas into the menagerie of astronomical objects visible today and the ninety-two natural elements we find around us? First it had some help from gravity. Although gravity is a very weak force, over many millennia it acted to pull hydrogen atoms closer together. Clumps of hydrogen formed and as they got bigger they exerted more of a gravitational pull, drawing in more hydrogen.

As these balls of hydrogen grew, the pressure on the gas in the centre of the ball increased because of the weight of all the hydrogen above it, and with this increasing pressure came higher temperature. (Think of inflating a bicycle tyre: the more you pump it up, the hotter it gets.) Higher temperature means that the atoms are moving at higher speeds and in the high-pressure core of a proto-star they collide against each other with increasing violence. At a certain temperature the collisions are so forceful that the atoms' outer electrons – which have a negative charge – are knocked away from their nuclei which, in the case of hydrogen, are made of just a single subatomic particle, a

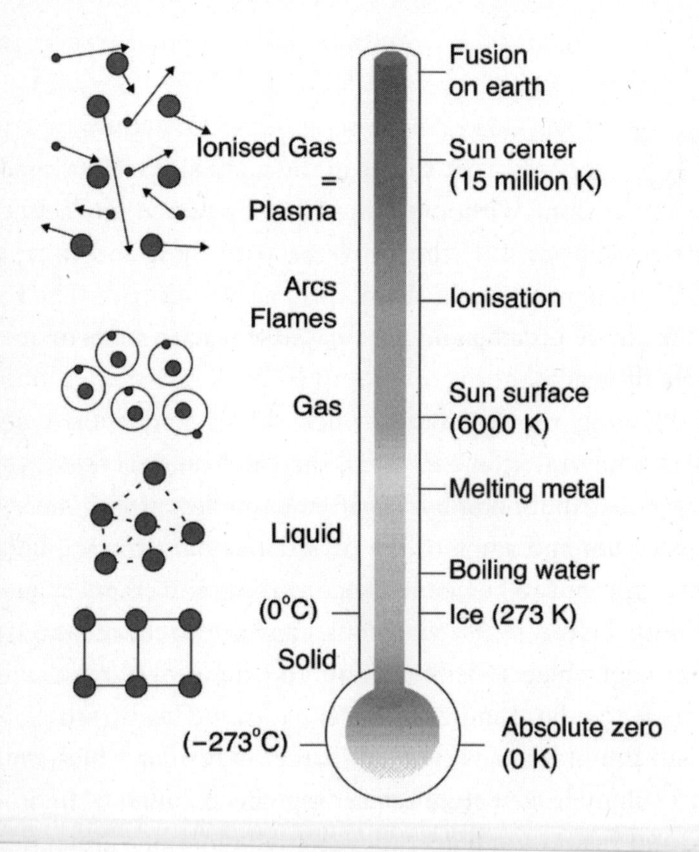


**At high temperatures, plasma – an ionised gas – represents a fourth state of matter after solids, liquids and gases.**

(Courtesy of CEA France)

positively charged proton. The result is a plasma; a hot maelstrom of charged particles.

Once the nascent star grows to a certain size – roughly 28,000 times the mass of the Earth – the temperature in its core reaches around 10 million °C and fusion starts. Fusion is simply the melding together of two nuclei to make a larger one but it's not an easy thing to do because all nuclei, such as the protons knocking around in the core of a star-to-be, have a positive electric charge and similar charges repel each other. When the temperature gets into the millions of °C, however, the nuclei are slamming together with such force that they get past the electric repulsion and are hooked by another short-range force, the one that holds protons and neutrons – their uncharged companions – together in a nucleus. The two colliding protons have to get within a subatomic arm's length before this attractive force can grab them and bind them together to make a new nucleus. But two protons don't make a very stable nucleus by themselves, so most of these pairs split apart again almost immediately.

Very, very occasionally one of these brief fusions is quickly followed by one of the protons decaying into a neutron. A nucleus made of a proton and a neutron – known as a deuteron – is very stable and so the new nucleus survives. Over time this process creates more and more deuterons in the heart of the proto-star and once there are enough of them other reactions start to happen. For example, one of the deuterons can fuse with another proton to produce helium-3 (two protons and one neutron) and, once there are enough of them, two helium-3s can fuse to form a helium-4 (two protons and two neutrons) with two protons left over. These reactions form the start of a chain of fusions which eventually also produces the elements lithium and beryllium.

These fusion reactions produce heat as a by-product because a nucleus of helium-3, for example, is slightly less heavy than the pair of reacting nuclei that created it – a deuteron and a proton in this case. This mass isn't lost; it is converted into energy during the fusion. So once this chain of reactions gets going, and untold numbers of nuclei are fusing, the heart of the proto-star becomes a raging furnace, further raising the temperature and causing more reactions. This simple process transforms the ball of gas into a fully-fledged star and it – or a very similar reaction chain – is what powers all stars, from the very first which are thought to have ignited about 150 million years after the Big Bang and throughout the 13.7 billion-year history of the Universe.

Fusion has more tricks too. Towards the end of a star's life, when it has burned up all its hydrogen, it starts to consume helium in a reaction chain that can produce beryllium, carbon and oxygen. When all the helium is used up, other chains begin that consume those nuclei to make even heavier ones. In this way, in the dying days of a star, all of the elements up to iron are created by fusion. Finally, when no fusion fuel remains, the remnants of the star collapse under their own gravity. If it is a large star, this collapse will release so much gravitational energy that it would blast the outer layers of the star outwards in a cataclysmic explosion, a supernova. The energy of a supernova is so intense that it causes further fusions in the heavy nuclei remaining in the star's ash. These fusions produce all the heaviest elements from iron up to uranium and beyond.

So, over the lifetime of a star, fusion takes the raw material hydrogen and forges it into all the other elements of the periodic table. And when the star explodes at its end, it spreads those elements out into space where they mix with fresh hydrogen and then slowly coalesce into new stars and planets. So these second-generation stars and the accompanying planets that form around them contain a mixture of elements, allowing some of the planets to form rocky surfaces, oceans, atmospheres and life. Every atom in your body, apart from the hydrogen, was created by fusion in a long-dead star.

* * *

Scientists spent the second half of the nineteenth century and the early part of the twentieth figuring out what made the Sun and all the stars shine. It was a mystery to them how the Sun could pump out such prodigious amounts of energy for billions of years without running short of fuel. By the late 1930s they had worked out the rough details of the fusion reactions described above and had their answer. That answer planted the seed of an idea into a number of minds. The seed wasn't able to grow for a while because of World War II but once that was finished it soon began to sprout. What was that seed? It was the idea that if fusion can power the sun for billions of years, could it supply similarly endless energy on Earth, if it could be mastered? The ancient Greek mythological figure Prometheus stole fire from the gods and gave it to humans, leading to progress, technology and civilisation. Could science steal the power of the Sun and rekindle it on Earth for the good of all humankind?

Prometheus came to a sticky end – chained for eternity to a rock for his crime. Postwar scientists didn't have a vengeful Zeus to worry about and, in fact, thought that taming fusion was going to be relatively easy. Stars make it look easy: lump together enough hydrogen, add gravity and fusion just . . . happens. On Earth they didn't have some of the benefits that stars enjoyed, including the weight equivalent to many thousands of Earths pressing down on the core to heat and compress hydrogen to fusion temperatures. Scientists would have to find some other way to heat and compress hydrogen – how hard could it be?

Although the war years had been devastating, they had produced some technological wonders. At the start, some men had still fought on horseback but the war was soon all about fast-moving armoured tanks, long-distance aerial bombardment, vast aircraft carriers and submarines. By the end of the fighting there were rockets able to hit targets hundreds of miles away, planes with jet engines and, ultimately, a bomb able to destroy a whole city. For some scientists after the war there was a sense of optimism. If they could achieve so much in six years of war, imagine what they would be able to do in peacetime.

One of those things was to develop nuclear power. But this was not fusion, it was the other sort of nuclear reaction, fission, the process behind the atomic bombs dropped on Hiroshima and Nagasaki. Fission is, in a sense, the opposite of fusion. In fission some of the very largest nuclei known, such as uranium, are split apart into two new nuclei. The starting nucleus is slightly heavier than the fragments that result from the fission and this missing mass is converted into energy during the process. Unlike fusion, fission doesn't need high temperature to take place: the large nucleus will split apart easily if hit by a fast-moving neutron. It was the discovery in 1938 that when some nuclei split they produce neutrons as a by-product which led to the realisation that you could start a chain reaction: one nucleus is hit by a neutron; it splits and spits out two neutrons; these hit two more nuclei which split to produce four neutrons, and so on. This chain reaction is what makes an atomic bomb – a fission bomb – possible: if you bring together a lump of one of these so-called fissile materials – such as uranium or plutonium – larger than a certain critical size, the chain reaction will start spontaneously and run out of control, exploding in a split second.

Before the scientists involved in the Manhattan Project – the Allies' wartime project to develop an atomic bomb – built an explosive, they tested the chain reaction in a controlled way, in the world's first nuclear reactor. This was built, in secret, at the University of Chicago in a squash court underneath the stands of its sports stadium, Stagg Field. Known as Chicago Pile-1 because it was a pile of uranium and graphite blocks, its construction was supervised by Enrico Fermi, a celebrated Italian-American physicist. Graphite absorbs neutrons and so slows the reaction. The blocks in the pile were carefully arranged so that there was enough uranium placed close enough together to sustain a chain reaction, but not quite enough for it to run away and explode. There was no radiation screening around the reactor and no protection from possible blasts – Fermi was sufficiently confident of his calculations to decide that they were unnecessary. In the mid afternoon of 2nd December, 1942 one of Fermi's assistants slowly pulled out a graphite control rod from the centre of the reactor. This reduction in the amount of graphite in the reactor was calculated to be just enough to allow the chain reaction to get going. Fermi watched a neutron counter and saw the number of neutrons swell as the rod was extracted. With a group of dignitaries looking on, Fermi ran the first controlled nuclear reaction for almost half an hour and then reinserted the rod to shut it down.

After the end of the war, engineers didn't waste any time putting this new technology to commercial use. The first nuclear reactor to produce electricity was built in 1951 in the United States. The first to supply power to the grid was in the Soviet Union in 1954. And the first truly commercial nuclear power plant began work in the United Kingdom in 1956. At the time, many expected nuclear power to produce electricity that was cheap and limitless. But there are problems with using fission as an energy source. First, uranium is a finite resource and some predictions say it may get scarce before the end of the twenty-first century. Then there is safety: because fission reactors rely on a chain reaction, it is possible for the reactions to run away too quickly and for the reactor to overheat. A reactor cannot cause a nuclear explosion like a atomic bomb because the fissile material inside is too spread out, but it can overheat and melt the core – as happened at Three Mile Island in 1979 – or catch fire – as at Chernobyl in 1986. Reactors contain many tonnes of uranium or plutonium fuel and, if they have been running for a while, also a lot of radioactive spent fuel, some of which is extremely harmful to people. When accidents happen, the danger is that this radioactive material will spread far and wide. At Three Mile Island the material was contained; at Chernobyl it wasn't.

Thirdly, there is the problem of waste. A typical nuclear power plant generating 1,000 megawatts (1 GW) of power produces around 300 cubic metres ($m^3$) of low- and intermediate-level waste per year and 30 tonnes of high-level waste. These quantities are tiny compared to the waste – some fairly toxic – from a coal-fired power station of similar output. But nuclear waste, parts of which can remain radioactive for hundreds of thousands if not millions of years, is more of a problem to deal with. Low- and intermediate-level waste can be buried close to the surface and its radioactivity will decline to safe levels in a matter of decades. High-level waste needs to be disposed of in such a way that it will remain inaccessible to humans or any other species for tens of thousands of years. Just imagining how to do that is a tall order and only a few countries have got to grips with the problem by building permanent repositories deep underground that will be sealed when full. Other countries keep their waste in carefully guarded facilities on the surface. All of the world's reactors combined produce a total of 10,000$m^3$ of high-level waste per year.

In the tumultuous postwar world, as a few countries raced to commercialise nuclear (fission) energy, some scientists realised that it would be a much better idea to try to generate power using nuclear fusion. The arguments in favour of fusion are compelling. First there is the fuel: fusion runs on hydrogen or, more correctly, on two isotopes of hydrogen known as deuterium and tritium. Deuterium is simply a deuteron (a proton and neutron) with an electron added; tritium has two neutrons in its nucleus. If you fuse deuterium and tritium you get helium and a neutron.

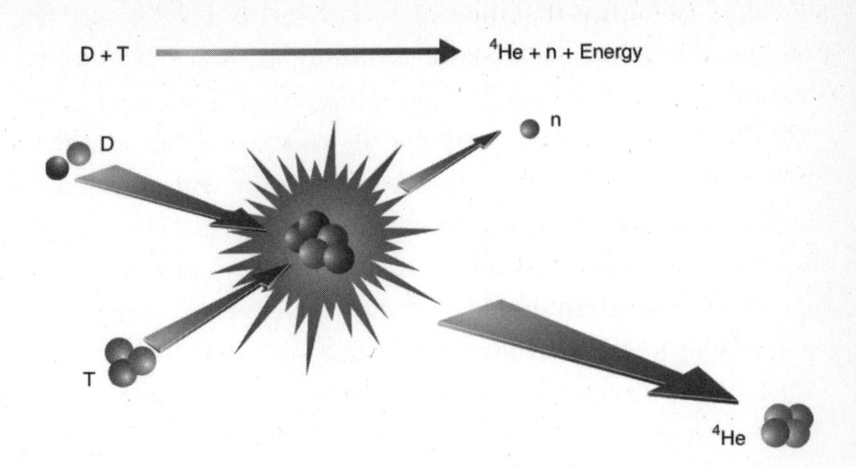


**Slam them together hard enough and a deuteron (D) and a tritium nucleus (T) will fuse, producing helium ($^4$He), a neutron (n) and lots of energy.** (Courtesy of EFDA JET)

Deuterium can be easily extracted from water. One in every 6,700 atoms of hydrogen in seawater is a deuterium atom. That doesn't seem like much but given the amount of water in the world's oceans there is enough deuterium there to supply all the world's energy needs for billions of years. Tritium is trickier because it is an unstable nucleus with a half-life of twelve years and so it would have to be manufactured. The easiest way to do this is with lithium, a metal used in some batteries. Lithium, when bombarded with neutrons, splits into helium and tritium. Any source of neutrons can cause this reaction and, since fusion reactors themselves are prodigious producers of neutrons, it is thought that a portion of a reactor's neutron output could be devoted to producing tritium fuel for its own consumption. Lithium can be extracted from easily mined minerals and there is enough around to supply the world with power for several hundred years. When that runs out, there is enough lithium in seawater for several million more years.

This seemingly vast oversupply of fuel is understandable when you consider how little fuel a fusion reactor actually needs. A 1-GW coal-fired power station requires 10,000 tonnes of coal – 100 rail wagon loads – every day. By contrast, a fusion power plant with a similar output would have a daily consumption of just 1 kilogram of deuterium-tritium fuel. The lithium from a single laptop battery and the deuterium from 45 litres of water could generate enough electricity using fusion to supply an average UK consumer's energy needs for thirty years.

Fusion is a nuclear process, so some might be concerned about its safety. There are safety issues associated with fusion, but they are small compared to a fission reactor. It is actually very hard to keep a fusion reaction going and if there is any malfunction in the controls of a fusion reactor the process would naturally just stop. Even if the process did want to run away it couldn't do so for long because there is very little fusion fuel in the reactor. Unlike a fission reactor, which has years' worth of fuel in place in its core, the fuel in a fusion reactor at any one time weighs about as much as ten postage stamps and could keep the reactor running for only a few seconds. The fuel stored outside the reactor is at no risk of reacting: it will only start to burn when heated to more than 100 million °C in the reactor.

Tritium is a radioactive gas, so is harmful to people, but as it will be generated on site, a fusion power plant won't keep a large supply sitting around. In the unlikely event that, for example, terrorists blew up a fusion reactor or crashed a plane into it, or even if an earthquake and tsunami hit the plant as happened at Fukushima, the amount of tritium released would not require any evacuation of nearby residents. In any event, tritium is a form of hydrogen, a buoyant gas once used in balloons and airships; its natural tendency is to drift straight up.

A fusion reactor does produce some radioactive waste but,

again, the amount is tiny compared to a fission plant. The 'ash' of fusion burning is helium, the harmless inert gas that is used to fill party balloons, lift modern-day airships, and cool NMR machines. The metal and other materials in the structure of a fusion reactor, after decades of being bombarded by high-energy neutrons from the reactions, will become mildly radioactive. So when a plant is dismantled it will need to be buried in shallow pits for a few decades, by which time it would be safe to recycle. There is none of the high-level waste that takes millennia to cool down.

Fusion seems too good to be true and to the fusion pioneers in the late 1940s and early 1950s, although they wouldn't have known all of these details, it was clear that fusion would be a vastly superior energy source compared to fission. There was a certain idealism to these early followers of fusion, who were almost all in the United Kingdom, the United States and the Soviet Union. All physicists had been shaken by the power unleashed in the Manhattan Project and many felt a sense of responsibility for the devastation that the atomic bombs caused. Fusion provided a way to use nuclear technology peacefully, for the benefit of everyone. Much of the early work on fusion was done in weapons laboratories because that's where the nuclear physicists were, but many of them left the labs to pursue fusion outside the military complex.

With the technological optimism of the time, the early pioneers expected that they would be able to master fusion in about a decade and then move on to commercial power stations – a similar timeline to fission. They knew that they would have to get their hydrogen fuel very, very hot, at least 100 million °C. At such temperatures solids, liquids and even gases cannot exist, so they would have to deal with plasma – that fourth state of matter that exists in the Sun's core where negatively-charged electrons and positive ions move around independently. In the middle of the last century scientists didn't know much about plasmas, especially very hot ones, and they had to come up with a system that could contain the plasma, heat it hotter than the core of the Sun and then hold it there without it touching the sides, because its extreme temperature would burn or melt almost any material.

Undaunted by these hurdles, the early fusion enthusiasts exploited the key difference between plasma and normal gas: that it is made up of charged particles. When charged particles move around in an electric or magnetic field they feel a force pushing them in a particular direction. So researchers started building containers pierced by complicated magnetic and electric fields to push the particles of the plasma towards the centre and away from the walls. Sometimes these were straight tubes, sometimes ring-shaped doughnuts and other shapes. At first they were made of glass – the scientist's favourite building material – small enough to sit on a lab bench and sprouting an impenetrable tangle of wires, pumps and measuring apparatuses.

Soon researchers worked out how to create plasmas in their devices and how to heat them to high temperature, if only for a fleeting fraction of a second. No fusion yet, but success in containing and heating plasmas encouraged them to try out new ideas, build more devices and build them bigger. Their makers gave them strange names such as pinches, mirror machines, stellarators and tokamaks. One of the reasons they were not getting to fusion temperatures was that too much heat was escaping from the plasma, so they guessed that bigger was better, since it would take the heat longer to escape from the core of the plasma if there was more of it. Soon their machines were too big for lab benches and were taking up whole rooms, then filling large hangar-like buildings.

They encountered other problems too. Using high-speed cameras to observe the plasma – it glows, just as the plasma in fluorescent lights does – they saw it wriggling and bulging as if trying to break free of its bonds. These phenomena, known as instabilities, had never been seen before, perhaps because no one had tried doing these things to plasma before. To work out how to prevent them, the researchers had to adapt or make up the theory of plasmas as they went along.

A pattern began to emerge in fusion research: scientists would build a new machine; when it was working they would make progress towards fusion conditions but not quite as much as they had predicted; this could be because the machine underperformed or they encountered some new unforeseen type of instability; the way forward was to build another bigger and better machine, and so on. Fusion got a reputation for promising a lot but never delivering. The oft-repeated joke was, 'Fusion is the energy of the future, and always will be.'

By the 1980s the reactors had come a long way from the bench-top devices of the early days. During that decade the biggest fusion reactors built to date were completed: the Joint European Torus or JET, which is the size of a three-storey house, and its US counterpart, the Tokamak Fusion Test Reactor. These were meant to be the reactors that finally made it to the first great milestone of fusion: break-even. This is the situation when the power given off by the fusion reactions is equal to the power used to heat up the plasma. Thus far all reactors had been net consumers of energy. If fusion was going to be viable as a source of power it had to get over this hurdle. But despite the heroic efforts of researchers over more than a decade, neither of these great machines managed to get to break-even. JET's best shot, made in 1997, produced 16 megawatts of fusion power but this was only around 70% of the power pumped in to heat the plasma – nearly there, but not quite.

* * *

Some people have spent their whole working lives researching fusion and then retired feeling bitter at what they see as a wasted career. But that hasn't stopped new recruits joining the effort every year: optimistic young graduates keen to get to grips with a complicated scientific problem that has real implications for the world. Their numbers have been increasing in recent years, perhaps motivated by two factors: there is a new machine under construction, a huge global effort that may finally show that fusion can be a net producer of energy; and the need for fusion has never been greater, considering the twin threats of dwindling oil supplies and climate change.

The new machine is the International Thermonuclear Experimental Reactor, or simply ITER (pronounced 'eater') as it now likes to be called. Many machines over the past sixty years have been billed as 'the one' that will make the big breakthrough, only to stumble before getting there. But considering how close JET, its direct predecessor, got to break-even, ITER has to have a good chance. Those earlier machines were almost invariably built in haste, part of a breakneck research programme. ITER, in contrast, was in development for a quarter of a century before construction began and, because of the delicate politics of building an international collaboration, was subjected to endless reviews, reappraisals, rethinks and redesigns. It may not be the perfect fusion reactor but it is the best guess of the thousands of researchers who have contributed to its design since the mid 1980s.

ITER is not a power station; it won't be connected to the grid and won't even generate any electricity, but its designers are aiming to go far beyond break-even and spark enough fusion reactions to produce ten times as much heat as that pumped in to make it work. To get there is requiring a reactor of epic proportions. The building containing the reactor will be 60m tall and extend 13m underground – altogether taller than the Arc de Triomphe. The reactor inside will weigh 23,000 tonnes – continuing the

Parisian theme, that's more than three Eiffel Towers. The heart of the reactor, the empty space were the hot plasma will hopefully burn, is about four times the height of an adult and has a volume of 840m$^3$ – dwarfing JET's 100m$^3$.

At the time of writing, workers at the ITER site in Cadarache, in southern France, are laying foundations, erecting buildings, installing cables and generally preparing the ground. In factories around the world the various components that will make up the reactor are being built, ready to be shipped to France and assembled on site. The scale and the quantities are prodigious. In six different ITER member countries factories are churning out niobium-tin superconducting wires for the reactor's magnets. When finished, they will have made 80,000km of wire, enough to wrap around the equator twice. The giant D-shaped coils of wire that are the electromagnets used to contain the plasma are each 14m tall and weigh 360 tonnes, as much as a fully laden jumbo jet. ITER needs eighteen of these magnets. Perhaps the most mind-boggling statistic about ITER, and one of the reasons it is being built by an international collaboration, is its cost: somewhere between ⇔13 billion and ⇔16 billion. That makes it the most expensive science experiment ever built – twice as expensive as the Large Hadron Collider at CERN. The European Union, as the host, is footing 45% of the bill; the rest is being split equally between China, India, Japan, Russia, South Korea and the United States. According to the current schedule, the reactor will be finished in 2019 or 2020.

That huge sum of money is, for the nations involved, a gamble against a future in which access to energy will become an issue of national security. Most agree that oil production is going to decline sharply during this century. There is still plenty of coal around but burning it in large quantities increases the risk of catastrophic climate change. That doesn't leave many options for the world's future energy supplies. Conventional nuclear power makes people uneasy for many reasons, including safety, the problems of disposing of waste, nuclear proliferation and terrorism. The disaster at the Fukushima Daiichi nuclear plant in Japan following the earthquake and tsunami in March 2011 served to remind the world how even the most secure installation can still be vulnerable.

Alternative energy sources such as wind, wave and solar power will undoubtedly be a part of our energy future. They are, in a sense, just harvesting energy from our local giant fusion reactor, the Sun. The cost of electricity from alternative sources is high but has declined substantially in recent decades and with continuing improvements in technology it will come down further. It would be very hard, however, for our modern energy-hungry society to function on alternative energy alone because it is naturally intermittent – sometimes the sun doesn't shine and the wind doesn't blow – and also diffuse – alternative technologies take up a lot of space to produce not very much power. There are ways around the intermittent nature of alternative energy, such as energy storage and backup generators, but these all add to the cost. Finding enough space for wind and solar farms is more problematic, especially since the world's wide open spaces tend to be far from big cities where the bulk of energy is consumed and transmitting electricity over long distances causes sizable losses. The United Kingdom, as an example, has pushed heavily to build up its wind energy production and by 2011 had around 300 wind farms with a total of almost 3,500 turbines. But wind energy contributed only around 5% of the UK's total electricity production and, given how hard it often is to overcome local opposition to building new wind farms, it is not clear how wind energy could provide a substantial fraction of Britain's needs.

Difficult choices lie ahead over energy and, some fear, wars will be fought in coming decades over access to energy resources, especially as the vast populations of countries such as China and India increase in prosperity and demand more energy. Anywhere that oil is produced or transported – the Strait of Hormuz, the South China Sea, the Caspian Sea, the Arctic – could be a flashpoint. Supporting fusion is like backing a long shot: it may not come through, but if it does it will pay back handsomely. No one is promising that fusion energy will be cheap; reactors are expensive things to build and operate. But in a fusion-powered world geopolitics would no longer be dominated by the oil industry, so no more oil embargoes, no wild swings in the price of crude and no more worrying that Russia will turn off the tap on its gas pipelines.

Much expectation rides on the back of ITER but it may not even be the machine to make the breakthrough first. There is a rival fusion reactor technology that has been developed largely in nuclear weapons labs to help study the physics of nuclear explosions and it, rather than ITER, could reach break-even first. This type of reactor doesn't confine a large volume of plasma and heat it to fusion temperatures. Instead it takes a tiny capsule no bigger than a peppercorn, filled with deuterium and tritium, and crushes it to a density a thousand times that of lead using the highest energy lasers in the world. Assuming that the compression is clean and symmetrical, the extreme temperature and pressure created will spark an explosive fusion reaction like a tiny hydrogen bomb. Each explosion produces a relatively small amount of energy, but if the process can be coaxed to give a good energy yield and a plant can be created that produces ten or more such explosions per second, then you might have a power station.

The foremost machine attempting this type of fusion is the National Ignition Facility or NIF near San Francisco. This $3.5-billion machine was completed in 2009 and, at the time of writing, researchers there are still fine-tuning it to get the best possible compression of the capsules. Their aim is to achieve 'ignition,' a burning plasma that sustains itself with its own heat and generates more energy than was put into compressing it. If NIF researchers do succeed they predict that, using existing technology, they could build a prototype power plant in just twelve years. Not everyone in this branch of fusion thinks that is possible but they all passionately believe that such machines will one day provide a viable alternative to magnetic fusion machines such as ITER.

There are other avenues too; other reactor designs that were put to one side in the headlong rush to build the one, big machine. Rather than using magnets or lasers to manipulate a plasma, these use things such as heavy ion beams, extreme electrical currents, or hydraulic rams. If ITER and NIF should fail, or not succeed well enough, then these bypassed technologies could have their day. Some of them have already been adopted by a new crop of start-up companies which, backed by venture capital, are trying to make a dash to fusion with small teams of dedicated researchers in secretive private laboratories.

There are still many sceptics who say that fusion will never supply a single kilowatt of power to the grid because there are just too many scientific and technological uncertainties. But their views will not dent the conviction of those who have dedicated their lives to the dream of fusion energy, enduring ups and downs, dead ends, false trails and minor breakthroughs. The story of fusion is not just one of scientists toiling away in laboratories in isolation. Military expediency, international politics and historical serendipity have all boosted and buffeted the progress of fusion research. Funding for the increasingly expensive machines that fusion requires has ebbed and flowed depending on the eagerness of governments to find alternative sources of energy: the Middle East oil embargo of the 1970s led to a huge boost in funding for fusion but by the 1980s, when oil was cheap again, research money was harder to find. Atomic espionage, superpower summits, hijackings by Palestinian terrorists and the Iraq War have all impacted on fusion's fortunes. What has kept it going is the unwavering belief among the scientists who have embraced the field that one day it *will* work. Fusion science is not about seeking knowledge for its own sake; it doesn't have the intellectual appeal of the Big Bang, black holes, the human genome or the hunt for the Higgs boson; it is about hammering away at a stubborn nut in the conviction that one day it will crack. There's unlikely to be a eureka moment but one day the operators of ITER, or some other reactor, will get their settings just right, the plasma will get hot, stay hot, and burn like a piece of the Sun.

CHAPTER 2

# Britain: Thonemann and the Pinch

NOBODY IS QUITE SURE WHO HAD THE IDEA FIRST. AFTER THE whirlwind of discovery that was physics in the 1920s and 1930s, all the theoretical building blocks for generating energy by thermonuclear fusion were at hand and someone was bound to put them together. As often happens in science, the same idea popped up in several disconnected places at around the same time.

Hans Bethe, an émigré physicist from Germany who fled when the Nazis came to power in 1933 and settled at Cornell University, remembers having a conversation in Washington during 1937 with fellow émigré Leo Szilard from Hungary. Bethe had that year written a landmark paper that finally nailed down the fusion processes in stars which produce energy, so he was well placed to address the problem of creating fusion. Fritz Houtermans was another German-born physicist who fled the Nazis, but in his case to Kharkov in the Soviet Union. He is believed to have been carrying out experiments in fusion in 1937. And Peter Thonemann, an undergraduate student at the University of Melbourne in Australia, remembers working out a basic plan for a fusion reactor in 1939.

Although everything was ready for the pursuit of fusion to begin, the Second World War soon had physicists thinking of other things. After Nazi troops invaded Poland, Szilard drafted a letter alerting the US government that a fission chain reaction could be used to make a devastating bomb and warning that German scientists were working in this area. He persuaded his old friend and colleague Albert Einstein to co-sign the letter and they delivered it to President Franklin Roosevelt. That letter led eventually to the herculean Manhattan Project to develop an atomic bomb before the Nazis did. Bethe joined the Manhattan Project, heading the theory division at the top-secret Los Alamos laboratory in the New Mexico desert where many of America's and Europe's best physicists spent the latter years of the war.

Thonemann also joined the war effort, taking a job at the Australian government's Munitions Supply Laboratories in 1940 and later leaving to join the research department of Amalgamated Wireless near Sydney. As the war approached its conclusion, he took up his studies again at Sydney University. In Sydney, his passion for the possibilities of fusion continued. His thesis topic was how to measure the density of electrons in a plasma. Thonemann talked endlessly about fusion and at home he melted window glass in the oven in an effort to make doughnut-shaped vessels for plasma experiments.

Thonemann came from a comfortable suburban background in Melbourne, with a stockbroker father, mother, two brothers and a sister. At university in Melbourne he had enjoyed tennis and skiing, while in Sydney he made his own surfboard to ride the waves back at Rye near his home town. He was entertaining in company and played the piano well. But he gave up this seemingly idyllic existence to go somewhere he thought would help him in his quest for fusion: Oxford University.

Stepping off a ship into the Britain of 1946 must have been a shock. Southern Australia was physically untouched by the war, so the bomb-damaged English cities, the rationing of food and clothes, and the general air of exhaustion must have made it seem an alien world. But Oxford did have what he was looking for: the Clarendon Laboratory, chock full of some of the biggest scientific names of the day – many of whom had only recently been released from war work – along with all the experimental apparatus and skilled technicians that he would need. Thonemann had been accepted to study for a doctorate in nuclear physics at a salary of £750 per year. He came armed with notebooks full of calculations describing the conditions necessary to achieve a fusion reaction. But his supervisor, Douglas Roaf, had other ideas and set Thonemann to work developing ion sources. The topics were related, however, and Thonemann was able to simultaneously carry out work on fusion 'under the counter' – the hunt was on.

The roots of the search for fusion stretch back a century before Thonemann began tinkering away in Oxford. At that time an argument developed between physicists, geologists and biologists over the age of the Sun. Physicists of the nineteenth century had made huge strides in understanding the world around them and, emboldened by their success, were applying their theories to ever grander problems. One pivotal achievement was the laws of thermodynamics, the principles that govern the behaviour of heat. According to the First Law of Thermodynamics, energy, or heat, cannot be created or destroyed but can only flow from one place to another and transform from one form to another. Thus the gravitational energy of a ball at the top of a slope is converted into kinetic energy as the ball rolls down the hill. Or the electrical energy of a current in a wire is converted into heat and light by a lightbulb.

Physicists found that such laws seemed to apply in every situation that they studied, and rightly concluded that they must be universal. But when these fearless physicists applied the First Law to the case of the Sun, this produced troubling conclusions. Scientists could estimate the energy being pumped out by our neighbourhood star by measuring the solar heat falling on a patch of the Earth's surface and then extrapolating from that to the heat on the inside of a sphere with a radius equal to the Earth-Sun distance. The amount of heat was phenomenal and that then begged the question: if heat cannot be created out of nothing, where is it all coming from? The best source of energy at the time was coal, but if you had a ball of coal the size of the Sun burning to produce heat at the rate scientists had calculated, it would be reduced to ash in around 3,000 years, far too short a time for the formation of the whole Solar System.

Two of the titans of mid-nineteenth century science had a different idea: German physiologist and physicist Herman von Helmholtz proposed in 1854 that the Sun's heat came from gravitational energy. As the Sun contracts this energy is transformed into heat and the body of the Sun glows hot, radiating light. Scottish physicist William Thompson, later Lord Kelvin, came to a similar conclusion and calculated that with this source of energy a body the size of the Sun could have been around for 30 million years.

This seemed a much more reasonable figure for the age of the Sun, but it didn't please everyone. Charles Darwin had published his new theory of Evolution in 1859 in his book *On the Origin of Species by Natural Selection* where he included a rough calculation of the age of the Earth made by studying erosion processes in the part of Kent where he lived, known as the Weald. His estimate was 300 million years, and he also concluded that Evolution would need this sort of time to produce the variety of life he saw around him. As the heat of the Sun was needed for life to exist, the Sun must be at least that age. Geologists, too, required an Earth aged in the hundreds of millions of years to explain the transformations of rock that they observed. The debate over the age of the Sun and Earth raged for decades and Darwin was sufficiently troubled by Thompson's argument that he removed any mention of timescales from his last editions of *On the Origin of Species*.

The solution to this mystery began to come together late in the century with a discovery entirely unrelated to astrophysics, geology or Evolution: radioactivity. French physicist Henri Becquerel first noticed that uranium salts, when he left them on top of a photographic plate wrapped in black paper, left an image of themselves when the plate was developed. The salts were emitting some sort of invisible radiation that could penetrate paper and expose the plate. Marie Curie and her husband Pierre, as well as others, continued the study of this phenomenon, which the Curies dubbed radioactivity, identifying different sorts of radiation and isolating two new radioactive elements, radium and polonium. Radium, in particular, was highly radioactive – more than a million times that of the same mass of uranium – and to those early pioneers it exhibited a fascinating property: it was hot, all the time, irrespective of the surrounding conditions. The metal appeared to be breaking the First Law of Thermodynamics. Where was all the heat coming from?

That question was answered in the following decade by Albert Einstein as a consequence of his theory of Special Relativity. His famous equation $E=mc^2$ broadened the First Law of Thermodynamics by including matter. Energy could seem to disappear if it is converted into matter, and similarly matter can be transformed into energy. Because the speed of light ($c$ in the equation) is a large number, a very small mass ($m$) of matter converts into a huge amount of energy ($E$).

Scientists soon realised the atoms discovered by Becquerel, the Curies and others are radioactive because they are unstable, so over time the nuclei of their atoms split apart into other, smaller nuclei. With each decay, a tiny bit of the nucleus' mass is converted into energy, explaining the heat produced by radium and the rays that were blackening photographic plates. What those researchers didn't know was that radiation could endanger your health. Marie Curie carried around samples in her pockets and kept them in her desk, enjoying looking at the blue-green light they gave off in the dark. She died in 1934 of aplastic anaemia, almost certainly as a result of exposure to radiation. Today, her notebooks and even her cookbook from the 1890s are considered too dangerous to handle without protective clothing and are kept in lead-lined boxes.

Scientists almost immediately began to wonder whether radioactivity was the source of the Sun's heat. But observations of the Sun showed that it didn't contain much radioactive material. It was mostly made of hydrogen, the smallest and lightest element which couldn't decay into anything smaller.

The decisive clue to the source of the Sun's energy was provided by the British chemist Francis Aston who, in 1920, was trying to prove the existence of isotopes, versions of an element that have different masses but identical chemical properties. Later he proved that isotopes do exist and, significantly, that their masses are always rough multiples of the mass of hydrogen. So the most common isotope of carbon has the mass of roughly twelve hydrogens, but there are also isotopes of carbon that weigh thirteen hydrogens and fourteen hydrogens. Although it wasn't known at the time, this is because atomic nuclei are made up of both protons and neutrons which have roughly the same mass. The normal hydrogen nucleus is just a single proton, while the carbon nucleus has six protons plus six, seven or eight neutrons. But in 1920, as part of his search for isotopes, Aston took a number of different elements and made very precise measurements of the masses of their atoms. As everyone expected, the mass of helium, the second-smallest nucleus, was around four times the mass of hydrogen. Aston's measurements were so accurate, however, that it was possible to conclude that while helium's mass was close to that of four hydrogens, it wasn't *exactly* the same – helium weighed slightly less than four hydrogens.

At that time, it was thought that helium really was made from four hydrogens, so the fact that the masses were slightly different was significant. One who spotted the importance of this result was Arthur Eddington of Cambridge University, one of the leading astrophysicists of the day. Eddington was an enthusiastic early advocate of the theory of Relativity and maintained contact with Einstein during the First World War when most British scientists shunned any contact with their German colleagues. Eddington was a committed pacifist and when he was called up for military service in 1918 he refused, risking prison. Prominent scientists rallied to his cause and the Astronomer Royal, Frank Watson Dyson, argued that his expertise was essential to an experiment they were to carry out to put Relativity to the test.

The scientists' pleas won the day and in 1919 Eddington and Dyson travelled to the island of Principe off the west coast of Africa to observe the solar eclipse of 29th May. One of the predictions of Einstein's General Theory of Relativity is that the gravity of a massive object would deflect the path of a beam of light. The object had to be very, very massive to observe this weak effect and Eddington and Dyson's aim was to use the Sun. With the Sun's light blocked out by the moon during the eclipse, it would be possible to see stars whose light passes close to the Sun. If gravity does indeed bend beams of light then, as the Sun moves across the sky, just before it obscures a star, the light from the star would appear to move as its path is curved by the Sun's gravity. The pair did see stars appear to move and when they revealed their results back in Britain the news was reported around the world as the first conclusive proof of the truth of Relativity. Eddington, and Einstein, became household names.

At that time, Eddington was also working on a theoretical

model of the interior conditions of stars, even though the source of their energy was still unknown. Some still adhered to Kelvin and Helmholtz's gravitational explanation but Eddington was convinced that some kind of nuclear process was more likely. As a result, he jumped on the measurements made by Aston and in August 1920, in an address to the British Association for the Advancement of Science, proposed a new theory. He suggested that in the searing heat at the centre of the Sun hydrogen atoms are fusing to form helium atoms and, if the loss in mass in this process measured by Aston is converted into energy, this could prove to be the Sun's energy source. Eddington estimated that if 5% of the sun's mass is hydrogen (we now know that it is actually around 75%) and if, according to Aston, 0.8% of the hydrogen's mass is converted into energy during fusion, then the Sun – at its current rate of heat production – will last about 15 billion years. He added, somewhat prophetically:

> If, indeed, the subatomic energy in the stars is being freely used to maintain their great furnaces, it seems to bring a little nearer to fulfilment our dream of controlling this latent power for the well-being of the human race – or for its suicide.

If the Sun had indeed burned for billions of years, that gave scientists – be they evolutionists, geologists or astrophysicists – all the time they needed.

Eddington continued to try to incorporate nuclear reactions into his theory of the interior of stars but it was still beset with problems. Although Aston had shown that combining four hydrogens to produce a helium freed up mass to convert into energy, no one knew how to make that process happen. And of the nuclear reactions that could be performed in the laboratory, none of them released enough energy to power the Sun.

Another worry was that, according to the classical physics that prevailed at the time, hydrogen nuclei just would not fuse. To react hydrogen it would be necessary to strip off its outer negatively-charged electron, leaving just the tiny exposed nucleus with its positive charge. For fusion, two such nuclei must slam into each other with such force that they get so close together that it is then more advantageous for them to merge than fly apart again. It's similar to what happens when two droplets of water are pushed together: at first they seem to try to stay separate as if surrounded by elastic skins, even though they are squashed up against each other, until eventually the best way to relieve the pressure is to merge into a single drop. The problem with two nuclei is that they carry the same positive electric charge and so repel each other – just as bringing the same poles of two magnets together creates a repulsive force – and the closer they get together, the stronger the repulsion. Classical physics predicts that it is virtually impossible to bring two nuclei close enough together to fuse.

In the 1920s, however, a new show arrived in town: quantum mechanics. In quantum mechanics there are fewer yes or no answers and more probabilities. Impossible things are allowed by quantum mechanics, they just have low probabilities of happening. A young Russian physicist called Georgii Gamow was in 1928 the first to apply quantum mechanics to nuclear reactions. He reasoned that it was not impossible for two nuclei to get close enough to fuse, and he developed a formula to find the probability of such a reaction.

Using Gamow's formula, Fritz Houtermans, then at the University of Göttingen in Germany, and Welsh-born astronomer Robert Atkinson began to look at what might happen to nuclei knocking around together under the sort of conditions that Eddington predicted would exist in the heart of the Sun. The two scientists complemented each other perfectly for this task: Houtermans was an experimental physicist who had worked with Gamow at Göttingen and knew about the application of quantum mechanics to the nucleus but not about the interior of the Sun; Atkinson knew all about Eddington's theory of the Sun but little about quantum mechanics. They calculated that under Eddington's predicted conditions there would be a healthy rate of reactions between colliding hydrogen nuclei. Their 1929 paper on the topic is thought by many to be the starting point for thermonuclear fusion energy research.

It was now time for experiments to take centre stage. During a visit to Cambridge, Gamow discussed his work on the quantum mechanics of nuclei with a young physicist at the Cavendish Laboratory there called John Cockcroft. This spurred Cockcroft to develop, along with his colleague Ernest Walton, a device for accelerating hydrogen nuclei or, as they were then becoming known, protons. Protons are a constituent part of all nuclei, along with neutrons. Hydrogen, the simplest nucleus, is made up of a single proton. Cockcroft believed that if Gamow was right, his accelerator would be able to accelerate protons to a high enough speed so that, if they collided with other nuclei, some fusions might take place.

By 1932, Cockcroft and Walton had built their accelerator and used it to fire protons at a sample of the metal lithium. They found that at relatively modest energies, the protons were able to penetrate into the lithium nuclei and split each one into two helium nuclei. This was hailed as a triumph at the time: the very first 'splitting of an atom'. Many years later, the pair would share a Nobel Prize for their achievement. It was also significant because this reaction was the first to produce a large amount of energy. The resulting pair of helium nuclei carried more than 100 times the energy of the proton that caused the reaction. The reaction was too difficult to achieve to be a practical source of energy for the Sun, but it at least showed that it was possible to liberate a lot of power from nuclei.

Back at the Cavendish Laboratory, a colleague of Cockcroft's called Mark Oliphant made some improvements to the design of the Cockcroft-Walton accelerator so that it could separate out and accelerate deuterium nuclei. Deuterium, with its extra neutron in the nucleus, is identical to hydrogen in every way except that it is twice as heavy. This similarity makes it very hard to distinguish from hydrogen – its existence had only been confirmed a couple of years earlier. Deuterium's discoverer, American physical chemist Gilbert Lewis, had only just managed in 1933 to separate out a usable quantity of so-called heavy water, made from oxygen and deuterium rather than hydrogen. As soon as he had enough, Lewis sent a sample over to Ernest Rutherford, the formidable director of the Cavendish Laboratory.

Rutherford was a towering figure in early twentieth-century physics, having earned a Nobel Prize in 1908 for the discovery that natural radioactivity was due to atoms disintegrating and that it produced two different sorts of radiation. In 1911 he overturned the prevailing 'plum pudding' model of the atom – that it was a positively charged ball peppered with negatively charged electrons – by proving that an atom has a tiny but dense nucleus and electrons orbiting around it, a description that still holds true today. Rutherford had a domineering personality, a very loud voice, and ran the Cavendish as his personal fiefdom. He had overseen the work of Cockcroft and Walton and now, with Oliphant, he was going to see what he could do with deuterium.

Rutherford and Oliphant fired deuterium nuclei, or deuterons, at lithium and a number of different elements to see what nuclear reactions they could cause. Eventually they collided deuterium with deuterium and found they produced two different reactions: one producing an isotope of helium known as helium-3 (two protons and a neutron) and the other an even heavier isotope of hydrogen (one proton and two neutrons) that would eventually be called tritium. Both of these reactions produced excess energy, roughly ten times that of the incoming deuteron. But Rutherford, for one, was not convinced that you could ever produce useful amounts of energy by this method because, although individual reactions produce energy, only around one in every 100 million accelerated protons or deuterons actually caused a reaction. So overall there was a huge energy loss. Rutherford famously said at the time:

> The energy produced by the breaking down of the atom is a very poor kind of thing. Anyone who expects a source of power from the transformation of these atoms is talking moonshine.

The reason why using an accelerator to produce fusion was so inefficient was because most of the accelerated protons or deuterons get tangled up with the electrons orbiting around the target nuclei, losing most of their energy before they reach their objective. In the heart of the Sun the situation is very different because it's a plasma: the hydrogen atoms are stripped of their electrons and the nuclei can collide directly with each other without having to fight their way past electrons first.

Gamow, by 1938, was in the United States having defected from Russia. He decided to do so in 1932 because of Stalin's repression but his early attempts to escape with his wife Lyubov Vokhminzeva, also a physicist, were unsuccessful. First they tried to paddle a kayak 250 kilometres across the Black Sea to Turkey but were foiled by bad weather. A later attempt to cross from Murmansk in northern Russia to Norway ended the same way. They eventually succeeded the following year, but in a far less adventurous fashion: Gamow got permission for them both to attend a physics conference in Belgium and they absconded from there. Installed at George Washington University in Washington, DC, Gamow decided that enough was then known about nuclear physics to launch a concerted effort to explain the workings of the Sun. He teamed up with another émigré physicist, Edward Teller, also at GWU at the time, and they concluded that deuterium fusion had to be the source of the Sun's heat. Gamow felt that the time was right for a conference to debate the topic.

The star of that spring conference was Bethe who, with two colleagues, had recently completed a series of three articles summarising all that was then known about nuclear physics – work that colleagues called Bethe's bible. Bethe arrived not having thought much about the Sun's energy but he soon latched onto Charles Critchfield, a former student of Gamow's, who just before the conference had proposed a series of nuclear reactions that could power the solar furnace. They worked together during the conference with Bethe helping Critchfield iron out some problems with the scheme. In their proton-proton chain, two protons collide first to produce a deuteron (one proton transforming into a neutron). The deuteron then fuses with another proton to create helium-3. And finally two helium-3s merge to create normal helium-4 plus two protons.

During the meeting, Bethe began working on another chain in which a carbon nucleus is bombarded by one proton after another, transforming it into a series of carbon, nitrogen and oxygen isotopes – hence its name, the CNO cycle – until eventually it spits out a helium-4 nucleus and returns to its original state. For six months after the conference, Bethe continued working on the problem and developed a coherent theory of energy production in stars which he published in a groundbreaking paper. The proton-proton chain was, it turned out, the dominant mechanism in smaller stars, including the Sun; larger stars favour the CNO cycle. This work would, in 1967, win Bethe the Nobel Prize for Physics, but this particular line of thought was soon put to one side as the Second World War loomed. Soon many of the world's top physicists would be co-opted into the Manhattan Project and would turn their minds to the search for an atomic bomb.

So it was, in Oxford's Clarendon Laboratory a decade later, that Thonemann knew he had to get deuterons to fuse if he was to have any chance of generating power. But the question was: how? Rutherford had shown that using a particle accelerator was hugely inefficient. Thonemann realised, as others soon did too, that the Sun already had the best idea. Simply heat up your fusion fuel: when you heat something up its constituent atoms move faster and if you keep heating it, eventually the atoms will be moving so fast that collisions will become fusions, releasing more heat to keep the process going and hopefully some to spare. When the Sun was forming billions of years ago from clouds of gas, that initial heating was provided by gravitational contraction – so, in a sense, Helmholtz and Kelvin had been right about the Sun's original source of heat. But once the core reached 15 million °C, fusion ignited and from then on the outward pressure of all the heat and light produced by fusion counteracts the gravity, an equilibrium is reached and the contraction stops. Our mild-mannered local star is performing a continual balancing act: the huge crushing weight of its mass (330,000 times that of the Earth) is held perfectly in place by the slow-burning thermonuclear reactor at its heart.

But recreating a piece of the Sun on Earth is far from easy because of the extreme temperatures needed. If any plasma at that temperature were to touch the container it resides in, that container would be instantly melted or vaporised. So Thonemann had to figure out how to contain his deuterium plasma in such a way that it did not touch anything. The answer to this puzzle lay in the unique properties of plasma.

Plasma is the fourth state of matter, after solids, liquids and gases. You can turn gas into a plasma by just heating it up: at a certain temperature collisions wrench electrons free from the gas atoms. You can also make plasma with an intense electric field, which pulls the negatively-charged electrons and positive nuclei in opposite directions, eventually ripping them apart. Most flames are plasmas, as are electric sparks, lightning bolts and the glowing gases inside fluorescent tubes and low-energy light bulbs. Plasma is, in fact, by far the most common state of matter in the Universe since all stars and most of the gas between the stars are plasma. Planets like ours are rare islands of electrical neutrality in a highly-charged Cosmos.

The most noticeable difference between plasma and normal gas is that, because it is made up of charged particles, plasma is affected by electric and magnetic fields. In an electric field, plasma ions will all flow in the direction of the field and all the electrons will flow against the field (a normal gas is unaffected). What you get is an electric current in the plasma, just like the current you get from electrons flowing along a wire.

The effect of a magnetic field on plasma is more subtle. Charged particles don't feel anything in a magnetic field if they are stationary or moving parallel to the field lines, but when they are on the move, cutting across the magnetic field, they will feel a force that is perpendicular to both their direction of travel and the field direction. So an electric current flowing along a wire from, say, west to east across the Earth's magnetic field running from south to north will feel a force pushing it vertically upwards. This phenomenon is crucial to devices such as electric motors and actuators.

So scientists had a tool that could push plasma around, but how to fashion that into a container that can hold a plasma without it touching the sides? The germ of an answer was planted when in the first few years of the twentieth century a bolt of lightning hit the chimney of the Hartley Vale Kerosene Refinery near Lithgow, New South Wales, Australia. A Mr G. H. Clark of the refinery was so puzzled by what happened to the chimney's lighting conductor that he sent it to J. A. Pollock, a physicist at the University of Sydney. Pollock called in a colleague, mechanical engineer S. H. Barraclough, to look at it. The short section of copper pipe appeared to have been crushed by some huge force but, as far as they knew, all that had happened to it was that a large current pulse had flowed down it from the lightning strike.

Pollock and Barraclough developed an explanation for what had crushed the tube. It was well known that an electric current flowing down a straight conductor generates a magnetic field with field lines that loop around the conductor. But if you have an electric current cutting across magnetic field lines – even if the field is created by the current – the electrons in the current will feel a force. In this case, with a straight current and a field looping around it, that force is directed inwards towards the centre of the conductor. During the Hartley Vale lightning strike, the pulse of current down the copper tube was so great that the inward force was enough to crush the copper pipe as if it were a toothpaste tube.

The phenomenon that Pollock and Barraclough discovered, soon dubbed the pinch effect, was for many years considered a scientific curiosity without much practical use. Forty years later, at that same Sydney University, Peter Thonemann learned about the pinch effect and began dreaming of fusion. He realised that if you get plasma to flow along a tube, and so produce a current, that current will generate a pinch effect and keep the plasma away from the walls of the tube. But what about the end of the tube? If it's closed, the plasma will just accumulate there; if it's open it will all drain out. Thonemann's solution, which occurred to other scientists at the time, was to bend the tube around into a doughnut-shaped ring so that the plasma can keep flowing round and round as long as necessary.

In Oxford, eager to turn his ideas into reality, Thonemann wrote to the director of the Clarendon lab, Frederick Lindemann, otherwise known as Lord Cherwell, asking for aparatus to carry out experiments directed towards fusion. This was no routine matter for the young physicist since Cherwell was a powerful and well-connected man, having been a confidante of and chief scientific adviser to Winston Churchill during the war. Cherwell asked Thonemann to present his ideas in a symposium of Clarendon staff. So in January 1947, Thonemann stood up and explained his ideas for controlled thermonuclear fusion to a high-powered audience of physicists. Few queried his calculations, although there were questions about how much radiation the fusion reactions would produce. 'You managed to stay on your horse,' Cherwell said to Thonemann afterwards.

So, now with Cherwell's approval, Thonemann directed the Clarendon's in-house glassblower to make him a ring of glass tubing – a shape that mathematicians call a torus – with a diameter of around 10-20cm. The first things that Thonemann had to figure out were how to initiate the plasma – converting a neutral gas into a plasma with an electric field – and how to make a plasma current flow around the torus. Getting the current to flow is crucial, because without a current there is no pinch, but how to do it? Here Thonemann would exploit a trick known as electromagnetic induction.

Just as a wire carrying a current, when it cuts across a magnetic field, feels a force, so the reverse is also true: when a changing magnetic field cuts across a wire, the electrons in it feel a force and so start to flow as a current. Similarly, a changing magnetic field cutting across a torus filled with plasma will push the plasma around the ring. Thonemann achieved this using an electromagnet whose field is carried through the centre of the torus using a ring of iron that loops through the torus like links in a chain. Such induction only works, however, if the magnetic field is changing, so an increasing electric current in the electromagnet will create an increasing magnet field in the iron ring which will in turn induce a growing current in the torus. But it's not feasible to keep increasing the current forever, so Thonemann used an alternating current which swings one way then the opposite way in quick repetition. With an alternating current applied to the electromagnet, the magnetic field – and hence the plasma current – is always changing, flowing first one way then the other, except during those brief instants when it changes direction.

Roaf, Thonemann's supervisor, acquired from somewhere an alternating current generator that had been used during the war for radar work. Thonemann quickly discovered that the alternating current alone wasn't enough to start the plasma: he had to use a static electric field to initiate it and create a conducting channel of plasma around the torus before induction could kick in and make the plasma flow. Thus began a series of meticulous studies to see how plasmas behave in magnetic fields. The physics of plasmas was an obscure and little-studied field at the time, so much of what he tried was completely new. He measured the basic conducting and magnetic properties of plasmas. He measured the strength of the pinch effect using plasmas in straight tubes. He learned that the current channel through the plasma in a torus had a tendency to expand outwards until it touched the outer wall of the torus and was extinguished.

Thonemann's work did not go unnoticed. In December 1947, John Cockcroft, the physicist who had split the atom in Cambridge fifteen years earlier, asked Cherwell to see what Thonemann was up to. Just the year before, Cockcroft had founded Britain's Atomic Energy Research Establishment (AERE) on a former RAF base at Harwell, just 25 kilometres from Oxford. Cockcroft met with Thonemann several times and a few months later AERE took over funding Thonemann's work, and provided him with two assistants.

By this time, Thonemann had demonstrated the pinch effect in plasmas experimentally. Now it was time to ramp up the power to produce a much stronger plasma current and show that he could squeeze the plasma enough to generate high temperatures. Thonemann and one of his new assistants, W. T. Cowhig, worked on the theory of plasma pinches during 1948 and tested their predictions of the pinch strength on mercury plasmas in a straight tube. Because of the higher power, they would need a torus made of something stronger than glass. They also needed something to counter the tendency of the plasma current to expand towards the outer wall of the torus. Thonemann's solution was to build a copper torus with some wires running along the inner surface of the outer wall. When the plasma current was flowing, Thonemann would pass a current flowing in the opposite direction through these wires. The oppositely flowing currents repel each other and Thonemann could use this force to keep the plasma current away from the wall.

The copper torus was built and ready to run by the summer of 1949 when Thonemann invited Cherwell and Cockcroft to come and see it in action. The torus had two small glass windows and when Thonemann powered it up you could clearly see a stable and brilliantly glowing plasma current channel in the middle of the torus tube. Cherwell and Cockcroft were clearly impressed and they began visiting Thonemann's lab every week, usually on a Saturday morning, to keep a close eye on his progress. What they hadn't told Thonemann was that he was not the only researcher in Britain chasing fusion.

In 1946, George Paget Thomson, a physics professor at Imperial College in London, filed a patent for a torus-shaped fusion reactor.

Thomson was at the heart of Britain's scientific establishment. His father was J. J. Thomson, a Cambridge University physicist who won a Nobel Prize for the discovery of the electron and whose name is attached to many other discoveries. The younger Thomson also won a Nobel, in 1937, for showing that electrons behaved like waves as well as like particles. Before the war he had studied plasmas alongside his father, and during the war years he focused on nuclear physics, advising the British government that a nuclear bomb was possible. With that sort of experience it was no surprise that he would begin to think about fusion.

Thomson discussed his fusion reactor idea with colleagues at Imperial and with Rudolf Peierls, a physicist from Birmingham University who had worked on the Manhattan Project during the war at Los Alamos and knew of discussions there about ways of containing a fusion plasma. Peierls was sceptical and pointed out some problems he saw with the scheme, prompting Thomson to make a few modifications. Thomson filed his patent in May 1946. It described a toroidal reactor which uses the pinch effect to contain plasma. It didn't say how the gas would be ionised nor specify how it would be made to flow around the torus, though several methods were suggested. A torus 3m across, the patent said, would be large enough to accelerate particles and achieve fusion.

Thomson was not able to do much with his idea because he was called to New York to advise the British delegation to the United Nations Atomic Energy Commission for most of 1946. But in January 1947, Cockcroft invited him to a meeting at Harwell about the possibility of setting up a fusion programme at the new AERE laboratory. More than a dozen physicists gathered for the meeting from Imperial, Birmingham, Oxford and Harwell, including Peierls and another Los Alamos veteran, Klaus Fuchs. Thomson described his reactor and the various ways he thought electrons could be driven around the torus. Peierls again expressed his doubts and Cockcroft, although interested, didn't think the time was yet right to build a large-scale experiment, as Thomson wanted. It was agreed at the meeting that the teams at Imperial and Birmingham should carry out further small lab experiments. Thomson set two of his students, Alan Ware and Stanley Cousins, the task of showing the pinch effect in a toroidal vessel.

But Thomson, convinced that his scheme was workable, kept up the pressure. In May he wrote to Lord Portal, the government's Controller of Atomic Energy, arguing that he had done all the work he could on his patent and to prove the concept it was time to build a larger experiment than could be contained in a university laboratory. Thomson suggested that it could be built at the new research labs set up by the company Associated Electrical Industries (AEI) at Aldermaston Court. AEI was only too keen to take on the work, and even volunteered to pay for it. But Portal inevitably consulted Cockroft and he insisted that the work remain under Harwell's control. At a meeting in October to discuss Thomson's AEI proposal, Cockcroft again quashed the idea of going straight to a large reactor.

It was soon after that meeting that Cockcroft learned about Thonemann's work at Oxford and the contrast between the two approaches would not have been lost on him. Thomson put his ideas down in a patent from the start, based on a somewhat hazy theoretical understanding, and was using all his high-level connections to move straight to a full-scale reactor. Thonemann, on the other hand, was slowly and methodically testing his ideas in the lab, working out what would work and what wouldn't. Cockcroft made sure Thonemann had funding and staff.

Interest in fusion was starting to grow. One of Cockcroft's deputies, H. W. B. Skinner, Harwell's head of general physics, was called to report to a government atomic energy committee in April 1948 on activities at Imperial, Oxford and Harwell. He pointed out that physicists still had a rather tenuous theoretical understanding of plasmas and was sceptical of Thomson's proposals for accelerating electrons around the torus, although he was keener on Thonemann's inductive method. Skinner rightly indentified the key problem of confining the plasma with magnetic fields. 'It is useless to do much further planning before this doubt is resolved,' he wrote.

It was not until the following year, 1949, that Thonemann produced a pinched plasma in his copper torus and Ware and Cousins at Imperial achieved a similar feat. The stage was set for an expanded fusion research programme but events outside the world of plasmas and magnetic fields were about to intervene.

On 2nd February, 1950, Klaus Fuchs was arrested and just a month later was convicted of passing atomic secrets to the Soviet Union. Fuchs was born in Germany and became a communist as a student. He fled the Nazis in 1933 and settled in Britain. When the war broke out he was initially interned as a German national but influential professors persuaded the authorities to release Fuchs and he took British citizenship. Peierls recruited Fuchs to Britain's atomic bomb project and the two of them were soon transferred to the Manhattan Project in the United States. Fuchs said later in his confession that after Germany invaded Russia in June 1941 and Russia became allies with the United States and Britain, he felt that the Soviet Union had a right to know what the western powers were doing in secret. Around that time, Soviet military intelligence made contact with Fuchs.

At Los Alamos, Fuchs helped work out how to implode the fission fuel in the first plutonium bomb and made numerous other contributions. He was present at the Trinity test site for the first atomic explosion in July 1945. The following year he returned to Britain to join the new AERE laboratory at Harwell. But later in 1946, US cryptanalysts, following years of effort, cracked the code used by Soviet intelligence agencies and discovered that spies had infiltrated the Manhattan Project. Decoding was still slow and difficult but the analysts eventually deciphered messages suggesting that one agent for the Soviets was a British nuclear scientist. It wasn't until 1949 that suspicion fell on Fuchs and after being challenged by an MI5 officer he made a full confession, describing in detail how he had passed details of the Manhattan Project to his Soviet handlers since 1942. Fuchs was imprisoned until 1959 and then emigrated to East Germany.

The Fuchs case caused a near hysterical clamping down of security at British atomic facilities. Fuchs had known all about the fusion research going on in Britain and this caused great concern for Cockcroft. Although the goal of the fusion research was the controlled release of energy for power generation, not bombs, a fusion reactor would in the process produce copious amounts of neutrons, and neutrons could be used to convert the non-fissile but abundant isotope uranium-238 into plutonium. Plutonium is the key to one type of atomic bomb and at that time it was in very short supply.

Until then, the fusion research at Oxford and Imperial had been carried out openly and the researchers had published their results in academic journals. Suddenly, Thonemann and his colleagues found themselves being questioned about the implications of their work. They argued vociferously against classifying their research but to no avail: Cockcroft put strict limits on what could be published. Anything that described work on high-temperature plasmas was automatically classified, as was anything that suggested they were working towards a thermonuclear reactor.

The need for greater secrecy made it simpler for Cockcroft to do what he already knew was inevitable: it was time for fusion research to step up a gear and move to a scale that was too big for university laboratories. He decided to move Thonemann and his team from Oxford to Harwell towards the end of 1950. Six months later, Ware and a colleague moved from Imperial to AEI at Aldermaston Court to continue their work.

Thonemann moved into Hangar 7 at the former airbase. In the hangar, fusion experiments were set up inside a cage of wire mesh, soon dubbed the Birdcage, which protected them from stray electric fields of the other large machines nearby. The team grew quickly with new recruits from the universities and other government labs. One of their first tasks was to build an electrical power supply and then test it on various tori made of copper and quartz, the latter so that the researchers could see the plasma.

In the early days in Hangar 7 it became clear that there were some serious problems with Thonemann's scheme. With the high frequency alternating current they were using to drive the plasma current around the torus there are many moments when the current stops to change direction. Ions were drifting during those brief moments and hitting the walls of the torus, causing the plasma to lose heat. The new power supply that they had built provided some improvement, but as they ramped up the power the problems got worse. After several years trying to work around the problem, one of the team's new recruits came up with a radical proposal. Bob Carruthers had worked during the war on radar and had helped develop pulsed power supplies, ones that ramp up the current from zero to a high value in a pulse going only in one direction. Such a pulse fed through the electromagnet that linked into the torus would produce just a single burst of pinch effect, instead of the rapid beats from an alternating current, but it was worth a try.

Carruthers and a few others borrowed some components from another experiment in Hangar 7, set up a bank of capacitors – short-term stores of electric charge that would provide the current – and cobbled together a small torus by welding together two glass U-bends. Despite the Heath-Robinson nature of the experiment, the results were astonishing: they produced pinched plasmas that lasted just one ten-thousandth of a second, but the plasmas were much better contained than those produced by alternating currents. Soon the whole team was focused on pulsed plasmas and such was the improvement that in January 1954 experiments with alternating currents were abandoned altogether.

Work began to scale up. The researchers built a series of larger tori, Mark I to Mark IV, that were 1m across, and these produced ever greater plasma currents. There was still a major fly in the ointment, however. When working with a glass torus, Carruthers and a colleague took pictures of the glowing plasma current and found that it wasn't forming a steady ring around the centre of the torus but was wriggling around the doughnut like a meandering river. This was the first time fusion scientists had encountered what they now call instabilities, in this case a kink instability. Over the following decades, as currents and power levels increased, physicists would uncover a whole zoo of instabilities and had to learn how to suppress each species. Back in the mid 1950s, they were perplexed.

It turned out that kink instabilities are a natural consequence of the pinch effect. If you think of a plasma current in a straight line, the magnetic field that it induces is like a series of rings around the current evenly spaced along it. Any slight kink in the current causes the rings on the outside of the curve to be spaced more widely apart and those on the inside more closely together. Magnetic field lines more closely packed together means a stronger magnetic field, so the force on the current that creates the pinch is unbalanced, pushing to accentuate the kink. What was needed was some sort of restoring force, pushing the kink back into line.

While the team puzzled over this, the success of the pulse transformer technique was leading to pressure for a next step to an even larger machine, one that could actually produce the temperatures necessary for fusion. Thonemann did the calculations and estimated that they would need a metal torus 3m across with the tube itself 1m in diameter. Cockcroft took the proposal to his bosses at the newly created UK Atomic Energy Authority (UKAEA) late in 1954 and they unanimously approved it. They budgeted £200,000 for the project, with the reactor itself costing £127,000.

Hangar 7 became a whirlwind of activity. Thonemann and Harwell theorists continued to hammer out the details of the design, which was completed in the spring of 1956 and a contract was then signed with the company Metropolitan-Vickers to build the machine. Nothing this big had ever been built for a fusion experiment before. The pulse transformer, the biggest that had ever been built in Britain, weighed 150 tonnes. At one point there was concern over whether Vickers would be able to get enough of the particular grade of high-quality steel needed for the transformer. Luckily, a strike in the US electrical industry meant large quantities suddenly came onto the market. Others at Harwell worked feverishly on new measuring techniques so that when the machine was working they could accurately assess the temperature of the plasma, the size of the current and the density of electrons in the plasma. In July 1955, the project was given the codename ZETA, for Zero Energy Thermonuclear Assembly – 'zero energy' because they did not expect it to produce surplus power.

The problem of kink instability was still a headache. Until, that was, another new recruit from the Clarendon, Roy Bickerton, suggested applying another magnetic field to the plasma, one going round the torus, parallel to the plasma current. Moving charged particles stick to magnetic field lines, spiralling round them in a corkscrew path. The field lines also have tension to them, like a rubber band, so when a kink starts to develop, this so-called toroidal field gets stretched and starts to pull the plasma back into line. Fortuitously, that correcting pull is stronger than the pinch-induced tendency to accentuate kinks once they start.

Producing a toroidal field was relatively easy: it just meant winding a wire around the torus and passing a current through it. Bickerton built a new glass torus and tested various types of winding and current values, and found that he could suppress kinks over a wide range of conditions. In 1956, the difficult decision was made to add such windings to the design of ZETA, which added significantly to its complexity and cost. Nevertheless, in August 1957 ZETA was finished, on time and on budget, ready to fuse some plasma.

Despite the scale of the project they were now working on, Hangar 7 still had the clubby atmosphere of a university physics department. Most of the scientists came from universities such as Oxford and Cambridge or from government labs where many of them had worked together during the war. Pipe-smoking was *de rigueur* and, as was the uniform of the day, scientists wore white lab coats and engineers brown ones. When the pressure was really on, Cockcroft, who lived on the Harwell site, would sometimes come to the hangar in the evenings with a crate of beer to give the team some light relief.

The team working on ZETA was also relatively cut off from the outside world. The scientists didn't like it but secrecy was strictly enforced, so they couldn't publish papers and get recognition for their work, they couldn't give talks about fusion at conferences and they couldn't even discuss their work with fellow scientists, family or friends. This suited Cockcroft just fine because he believed Britain had a lead in fusion technology and he wanted to keep it that way.

Britain had something to prove on the nuclear front. The Manhattan Project during the war had been a genuine collaboration between the United States, Britain and Canada. But in 1946, the US Congress passed the McMahon Act which prevented foreigners having access to American nuclear secrets and the collaboration ended. The British government then had to take stock: should it take on the vast expense of developing its own nuclear weapons or just leave the nuclear arms race to the Americans? Arguments raged behind closed government doors in London. In the end, the nuclear enthusiasts won, in part because of a desire for national prestige, but also because of the expected industrial importance of atomic energy. That decision quickly led to the creation of the Atomic Energy Research Establishment at Harwell and the Atomic Weapons Research Establishment at Aldermaston.

Britain exploded its first nuclear weapon in 1952, the third nation to do so. It also switched on the first nuclear power station – Calder Hall – to supply commercial quantities of power to the grid in 1956. Cockcroft hoped that his team in Hangar 7 would pull off a coup in nuclear fusion too. That opinion was reinforced in 1956 when he visited nuclear research facilities in the US. Although what he was allowed to see was restricted, he got the impression that the US was spending a lot on fusion but had yet to make much progress.

That same year, Harwell got a valuable, but not necessarily complete, insight into what Russian fusion researchers were doing. During April an official Russian delegation, led by Soviet premier Nikita Khrushchev, visited Britain. Among the party was Igor Kurchatov, the USSR's leading nuclear scientist and father of the Soviet A-bomb and H-bomb. His laboratory, the Institute of Atomic Energy in Moscow, was the home of Russia's growing fusion programme. Kurchatov contacted Cockcroft and asked if he could visit Harwell and deliver a scientific lecture. Staff from all departments at Harwell crowded into the lecture theatre to hear Kurchatov speak, along with colleagues from AEI and the atomic weapons lab at Aldermaston. Placed on every seat was a printed copy of the lecture, in Russian and English.

**The Soviets visit Harwell, April 1956. Igor Kurchatov is on the far right and Nikita Khrushchev at the front, left of centre.**
(Courtesy of UK Nuclear Decommissioning Authority)

Entitled *On the Possibility of Producing Thermonuclear Reactions in a Gas Discharge*, Kurchatov's speech seemed daringly open to an audience forbidden from discussing their work. Although he didn't reveal any details of exactly what Soviet scientists were working on, he did discuss the complexity of the problem and how hard it was to draw firm conclusions. In particular, he described the difficulty of determining whether or not the neutrons produced by a plasma were really the result of thermonuclear reactions – an issue that would soon come to haunt the Harwell researchers.

Russian researchers in 1952, Kurchatov said, had obtained neutrons from deuterium pinched in a straight tube, but after some investigations found they had properties that were inconsistent with their coming from thermonuclear reactions.

Both Cockcroft and Thonemann suspected the lecture was a fishing expedition: Kurchatov wanted to find out – from the questions asked after his talk – what progress the British scientists had made. But Cockcroft was ready for that. He had given all the scientists attending a list of topics they were *not* allowed to discuss during the question and answer session. Thonemann came away from the talk with the (incorrect) conclusion that the Russians were not yet experimenting with plasma in a torus. Cockcroft, however, realised that they wouldn't take long to catch up, judging by how quickly they developed nuclear bombs after the end of the war. He set about speeding up Britain's fusion research.

Meanwhile, cracks were beginning to show in the secrecy surrounding fusion research. In December 1953, US president Dwight D. Eisenhower made a speech to the UN General Assembly which later became known as the 'Atoms for Peace' speech. In it Eisenhower pledged to make nuclear technology that didn't have military uses freely available for the benefit of mankind. Whether his intention was quite as altruistic as it sounded historians are still debating, but it had profound implications for those toiling away in secret government labs on nuclear projects.

One result, a few years later, was the creation of the International Atomic Energy Agency, the UN nuclear watchdog which oversees civil nuclear power and tries to ensure material is not diverted into weapons production. Another outcome of the speech was the International Conference on the Peaceful Uses of Atomic Energy, which took place in Geneva in August 1955. For those who had laboured in secret through the war and the decade that followed it, the Geneva conference was an astonishing event. Previously they couldn't even tell their own families what they were doing; now they could show it to the world and hobnob with fellow researchers from other nations whose work they knew nothing about. Scientists from western countries even exchanged notes with their opposite numbers behind the Iron Curtain.

The focus of the 1955 conference was nuclear fission, which would shortly be impacting directly on people's lives as the first power stations came online. Fusion scientists didn't get to join in this spirit of openness: the possibility of using a fusion reactor to produce plutonium for bombs meant that governments weren't yet ready to let that genie out of the bottle. A passing remark at the Geneva meeting, however, did set the ball rolling for fusion declassification. The Indian physicist Homi Bhabha, president of the conference, said in his opening address:

> I venture to predict that a method will be found for liberating fusion energy in a controlled manner within the next two decades. When that happens the energy problem of the world will truly have been solved forever for the fuel will be as plentiful as the heavy hydrogen in the oceans.

That teaser led to feverish press speculation about the existence of secret fusion research projects. Soon a number of governments, including the US and British, admitted the existence of their programmes, but few other details were released. That lack of information only fuelled the media's desire to find out more.

Soon after Cockcroft's trip to America in 1956, the US Atomic Energy Commission (AEC) formally suggested collaboration between the two countries' fusion programmes. That didn't lead to a flood of information back and forth across the Atlantic but the scientists did begin to visit each other's labs and the two sides agreed to a common secrecy policy: neither would publish anything without the other's approval. The British scientists were still agitating for more openness but the AEC took a firm line.

Researchers were permitted to publish a few papers around this time which described the science of fusion in general terms but nothing about specific machines or future plans.

On 12th August, 1957, ZETA was fired up for the first time. For the first few days the researchers used plain hydrogen while they worked out the reactor's optimum operating conditions and then switched to using deuterium. On 30th August, their detectors started to register the production of neutrons, the tell-tale signal of fusion reactions taking place. It wasn't long before they were getting a million neutrons per pulse and Harwell was soon buzzing with excitement. Could they really have struck oil so quickly? But they didn't want to get carried away and be fooled by a false neutron signal as the Russians had been five years earlier. It was impossible to tell at that time whether the neutrons were from thermonuclear reactions. The team didn't even have a way of accurately measuring the plasma temperature to know if it was hot enough for fusion.

To produce power from fusion, it's vital to have the plasma uniformly heated to a sufficiently high temperature for fusion reactions to start happening all through the core of the plasma. What the Russians saw, and the team in Hangar 7 feared being fooled by, was some other effect that wouldn't make viable amounts of energy, such as the plasma touching the torus wall and kicking off neutrons, or some flaw in the magnetic field that caused a small part of the plasma to be accelerated to high speeds but leaving the rest too cool. So the researchers opted for caution and avoided making any public statements until they were sure they were seeing thermonuclear neutrons.

Cockcroft, however, let the excitement get the better of him. On 5th September, a Thursday, he wrote to Edwin Plowden, chairman of the Atomic Energy Authority, telling him about the neu-

**Britain goes for the big time: the ZETA reactor at Harwell.** (Courtesy of UK Nuclear Decommissioning Authority)

trons but saying he wasn't yet 100% sure they were thermonuclear. Plowden inferred from this that the probability was high, though not quite 100%, and he wrote to the Prime Minister, Harold

Macmillan, the following Monday to tell him so. But by then, Macmillan could read about it in the newspapers for himself.

Somehow, news of the existence of ZETA had leaked out to the press along with the suggestion that something important was afoot. The day after Cockcroft wrote to Plowden there was a session on thermonuclear fusion at the annual meeting of the British Association for the Advancement of Science in Dublin and reporters were there in force expecting a big announcement of results from ZETA. Even the Irish prime minister Éamon de Valera was in the audience. Cockcroft had lined up two speakers for the session: George Thomson and John Lawson, a theorist who had worked at Harwell but had recently moved on to other things. Cockcroft had given Lawson strict instructions not to give anything away. At a press conference following the talks the pair were grilled by the press about what was happening at Harwell. Many of the questions they declined to answer because of the secrecy rules they were bound to, which annoyed the eager reporters. Thomson did concede that he thought at least fifteen years would be needed to build a power-producing reactor.

The newspapers were full of stories about fusion the next day, many extrapolating wildly about the prospects of fusion power. The *Financial Times* reported that ZETA had been producing neutrons since mid August, though this hadn't been mentioned at the Dublin meeting. Others said that ZETA had reached temperatures of 2 million °C. The more fanciful reports distressed the researchers at Harwell and Thonemann argued that they had to make an official statement to set the story straight. Cockcroft agreed but his hands were tied by the agreement with the US; any official statement would have to be approved by them too. A draft press release was drawn up and then cabled to Washington.

The first reaction from the AEC was to say that they would make a parallel announcement at the same time. But after US fusion scientists expressed scepticism at the British results, the head of the AEC, Lewis Strauss, said any announcements should wait until the next Geneva Conference on the Peaceful Use of Atomic Energy, a whole year away. Plowden would have none of it and they agreed to wait at least until the next meeting of Project Sherwood, the codename for American's fusion programme, in mid October when visiting British scientists could explain their results. Plowden told Strauss he didn't think he could hold off the British press any longer than that.

The Harwell team knew they needed more evidence, both of the plasma temperature and the thermonuclear nature of the neutrons, but Thonemann and others suspected that the Americans had an ulterior motive: they wanted more time for their own experiments to bear fruit so it would not look like they had been beaten in the fusion race. Harwell had to accept the delay, so the team continued to run ZETA, gather more results and look for firmer evidence.

At the beginning of October, however, two momentous events occurred which would load ZETA with enormous political baggage. On 4th October the Soviet Union launched Sputnik, the world's first artificial satellite. Less than a week later, Pile 1, a nuclear fission reactor at Windscale in Cumbria, caught fire and spread radiation across the local area. Although the Windscale pile was designed to produce plutonium for nuclear weapons – not generate electricity – the fire and its threat to the population was a major blow to the clean high-tech image of nuclear power, and the public began to question its safety. The Atomic Energy Authority needed something to distract people from the fire. ZETA, with its cleaner and safer form of nuclear power, would do the job perfectly.

On the other side of the Atlantic, the United States was reeling in shock from the launch of Sputnik. The US had always assumed it had an unassailable lead in high technology and that the Soviet Union would forever be playing catch-up. In part this was because the US had captured all Nazi Germany's top rocket scientists at the end of the war and spirited them back to America. The conquest of space was theirs for the taking, so they thought. While the launch of Sputnik was greeted with wonder by many, for the US military it inspired terror. If the USSR could launch a satellite, it could in theory drop a nuclear weapon from space onto anywhere on US territory – something for which America had no defence. The US needed some new breakthrough to show that it was still a technological powerhouse. Fusion could provide that breakthrough but the press might portray it as a British triumph, further humiliating the US. As a result, the AEC continued to play for time.

Every day that passed was torment for the members of the Harwell team who were desperate to tell the world about their achievement; especially since inquisitive newspapers were publishing ever-more speculative stories about what they thought was going on. This particular period of history, the 1950s, is perhaps unique in that scientists – and especially physicists – were treated as heroes. Although the nuclear weapons dropped on Japan had inflicted awful damage, creating them was a tremendous technical achievement and they had brought the war to a swift end. In the postwar years physicists served up more wonders: rockets, jet planes, televisions and nuclear power. On TV and in films they were portrayed as noble knights in white coats, able to solve almost any problem. In the uncertain world of the Cold War it was good to have such people on your side. And it was into this atmosphere that the Harwell team aimed to launch their reactor, which promised clean and virtually limitless power at very little cost. They were surely unprepared for the impact their announcement would make.

At the Project Sherwood meeting in October it was agreed that both teams would publish papers describing their results and the text of a new press release was hammered out, but the AEC continued to stall, arguing that more evidence was needed. Newspaper articles were beginning to carry claims that Britain was ahead of the US in fusion technology, and that Harwell's 'triumph' was being suppressed because of US-imposed security rules. Questions were asked in Britain's House of Commons. Before anti-US sentiments got carried away, Strauss agreed to go ahead with publication but first sent a delegation of US fusion scientists to Harwell to see ZETA for themselves. Following their visit in December, the US scientists still argued for more delay but it was agreed that both sides would publish scientific reports in the journal *Nature* early in the following year.

So it was that the press was invited to Harwell on 23rd January, 1958 and the researchers opened up Hangar 7 to show ZETA to the world. Because of their nagging doubts about the origin of the neutrons, the researchers didn't put anything into the papers in *Nature* about how the neutrons might have been produced. The press release given out to reporters, however, suggested the neutrons were *probably* thermonuclear. Scenting that this was a key issue, reporters repeatedly asked the Harwell scientists about the neutrons but only got evasive answers. At the press conference that day Cockcroft was similarly bombarded with questions and eventually admitted that he was 90% certain that at least some of the neutrons were thermonuclear.

The next day, ZETA was the top story across the globe and Cockcroft's 90% certainty was reported in every story. 'The Mighty ZETA,' trumpeted the *Daily Mail*'s front page. The *News Chronicle* declared, 'Britain Unveils Her Sun.' Many described ZETA as 'Britain's Sputnik.' The Hangar 7 team were suddenly celebrities, with their pictures on the front pages alongside brief pen portraits, often focusing on their comparative youth. 'These are the names which will be linked with the controlling of the H-bomb in the same way that Rutherford, Cockcroft, Fermi and others are bracketed with atomic energy,' said the *Mail*, continuing, 'Tall, dark and bespectacled, Thonemann has dedicated all his working years to tapping nature's biggest bank of energy.'

The Italian press gave ZETA more coverage than it had Sputnik. French papers were some of the few that pondered the 10% possibility that Cockcroft was wrong. *The New York Times* reported the view of one researcher that fusion reactors could power spacecraft. Newsreel cameras recorded the open day at Hangar 7 and a hastily produced TV programme explained the breakthrough to Britain's viewing public. Despite the simultaneous papers published in *Nature*, few reports mentioned the work being done in the US – ZETA was a very British triumph.

ZETA and its creators continued to be feted in the months that followed, but researchers were still concerned about whether they really were seeing what they thought they were seeing. Other pinch-based machines were also starting to produce neutrons, including a torus built by Ware at AEI and America's whimsically named Perhapsatron. But US scientists continued to be sceptical about ZETA's results. They just didn't believe that it could be getting up to the temperature of 5 million °C that the Harwell team was claiming, and any less than that would not be hot enough to cause thermonuclear reactions.

The questionable neutrons were about to come under close scrutiny. At the ZETA press conference in January, Basil Rose, a nuclear physicist from another section at Harwell, managed to get in to find out what all the fuss was about. Rose was in charge of Harwell's cyclotron, a particle accelerator that shared Hangar 7 with ZETA. He quickly realised that finding out more about the neutrons was crucial, and his experiment had a detector that could measure accurately the energy and direction of neutrons. Frustratingly, he had just leant the detector, called a diffusion cloud chamber, to scientists at University College London but they hadn't started using it yet and Rose persuaded them to ship it back to Harwell.

**Peter Thonemann (left) receives an achievement award from Sir Edward Hutton (right). Sir John Cockcroft looks on.**
(Courtesy of UK Nuclear Decommissioning Authority)

If ZETA's plasma was at thermonuclear temperatures, then the deuterons would be bouncing around in random directions, colliding and fusing. The neutrons they emitted would therefore be flying out in all directions equally and with similar energies. When Rose hooked up his cloud chamber to ZETA and studied the neutrons, this was not what he found. The neutrons were mostly emitted in line with the axis of the plasma current and more strongly in one direction than the other. To prove the point,

Rose got the team to run ZETA 'backwards,' with the plasma current flowing in the opposite direction to normal. Sure enough, the preferential direction of the neutrons was reversed. The conclusion: ZETA's neutrons were definitely not created by thermonuclear fusion.

In mid May the team announced these results at a press conference in London and a month later *Nature* published the details. Press reaction was sober and relatively restrained. Perhaps newspapers were embarrassed by their own wild extrapolations a few months previously. The *Manchester Guardian* mused over whether the obsession with secrecy was to blame:

> In a huge research project like that revolving around ZETA the day-to-day rubbing of shoulders with scientists of other specialities is the best safeguard of sound analysis and interpretation ... So it will inevitably be asked whether things might not have gone differently if the members of the ZETA team had been allowed to talk freely and informally to other scientists.

Fusion scientists always maintain that ZETA was a success. It was, after all, the first machine to achieve a large, stable pinched plasma at high temperature. Scientists at Harwell continued to use it up until 1968, garnering much useful information. But in the public mind ZETA will always be remembered as a failure: British scientific hubris dashed upon the rocks of a problem more complex than foreseen. It's true that the Harwell team were impetuous in going public with their results when they had very unreliable information about the temperature of the plasma and the nature of the neutrons, but they are not solely to blame for the mess that ensued. The need for a success after the shock of Sputnik and the Windscale fire meant that Harwell was under enormous political pressure to produce the goods.

Britain was also hungry for something to restore its national pride. Although it had emerged as one of the victorious powers at the end of the war, it was struggling to hold onto its seat at the top table. Britain had had to scramble to catch up with the superpowers in atomic power and weapons. Its economy was in tatters (rationing had only ended in 1954) and its empire, which had once spanned the globe, was being rapidly dismantled. In 1956, Britain's adventure with the French to seize control of the Suez Canal following nationalisation by Egypt was humiliatingly squashed by a disapproving US. With so little to celebrate, the British public embraced the scientists who had given them a world lead in this wonderful new technology, and felt betrayed when it was taken from them again.

A few months after the climb-down over ZETA, British fusion scientists, along with colleagues from all over the world, gathered in Geneva for the second 'Atoms for Peace' conference which this time was focused on fusion. Just before it began all sides declared they would declassify their fusion research. The US and Soviet fusion programmes vied to outdo each other in displays of research activity. The US stand cost millions to put together and contained four real fusion machines. The Soviets put on a similar show. With the veil of secrecy lifted, Harwell's researchers could see that they wouldn't be able to keep up for long.

In 1960 Britain's fusion researchers began moving to a new purpose-built laboratory at Culham, another former airfield some 10 kilometres from Harwell. The plan for a bigger and better ZETA 2 was abandoned, however, and the emphasis shifted to smaller machines to gain a better understanding of how plasma works. Thonemann left the programme and, in a sense, the 'heroic age' of British fusion was at an end. It would be some decades before another big fusion reactor was built in the UK but, elsewhere, things were just hotting up.

CHAPTER 3

# United States: Spitzer and the Stellarator

WINDING THE CLOCK BACK TO 1951, WHEN THONEMANN was already installed at Harwell, the United States didn't even have a research programme into controlled nuclear fusion. Just after the war, some of the scores of scientists who had been holed up in the Los Alamos laboratory for the Manhattan Project starting turning their minds to other things, in particular how to build a more powerful nuclear bomb based on the fusion of hydrogen, the H-bomb or 'Super' as scientists called it at the time. Enrico Fermi, who had built the first ever nuclear fission reactor – Chicago Pile 1 – in 1942, came to the lab and gave a series of lectures on thermonuclear reactions. This got some of his audience thinking about controlling fusion for energy production. During those days at Los Alamos, Edward Teller would organise 'wild ideas' seminars and some of those were devoted to the problem of how to control fusion reactions. Briton James Tuck and Polish mathematician Stanislaw Ulam did some calculations on the possibility of accelerating beams of deuterium ions to high energy and colliding them to cause fusion. They even carried out some experiments but the effort fizzled out. Now that the wartime bomb project was finished, many scientists at Los

Alamos were drifting back to their prewar jobs. Teller, for example, returned to the University of Chicago and Tuck went back to the Clarendon Laboratory in Oxford.

Things changed following the explosion of the first Soviet atomic bomb in August 1949. US strategists had believed they had many more years before the Russians caught up with their nuclear programme. The blast at Semipalatinsk was a wake-up call. If the Soviets could produce a fission weapon in just a few years, an H-bomb based on fusion might soon follow. The US had to get there first, so President Truman ordered a crash programme to develop the H-bomb. Teller returned to Los Alamos in 1950 and many others joined the effort. At Princeton University in New Jersey a branch of the programme, known as Project Matterhorn, was set up to work on theoretical aspects of the H-bomb. One of those recruited to work on it was the astrophysicist Lyman Spitzer Jr. An astrophysicist might seem an odd choice but Spitzer was an expert on the interstellar medium, the thin clouds of gas and dust that occupy the space between stars. Interstellar gas is mostly a hydrogen plasma and since the designers of the H-bomb knew they would have to master hydrogen plasma in their device, Spitzer was their man.

In March 1951, before Spitzer started on Project Matterhorn, he was due to take a skiing holiday in Aspen, Colorado. But on the morning of his departure he had a phone call from his father telling him that he should buy a copy of *The New York Times*. The paper reported that the Argentine dictator Juan Perón had announced that his country had achieved controlled nuclear fusion and was developing the technology to generate electricity for the benefit of all mankind. Details were few at the time but it was revealed later that an Austrian physicist called Ronald Richter had persuaded Perón in 1948 that he could provide Argentina with inexhaustible energy through fusion. Perón was an enthusiast for all things German and, without consulting Argentine physicists, essentially wrote Richter a blank cheque and built him a laboratory on the island of Huemul in a remote part of western Argentina. Richter's plan was to use some form of magnetic field to confine a plasma and react deuterium and lithium. According to Perón's statement on 24th March, Richter's experiment, dubbed the thermotron, had produced particles and energy consistent with fusion.

Spitzer set off for Aspen but the news reports from Argentina had set his mind racing. If you wanted to, how would you achieve a controlled fusion reaction? Spitzer was a gifted scientist who had studied at Yale, Princeton and Cambridge (under Eddington) in the 1930s. He played a key role in the development of sonar during the war and in 1947, aged 33, he was made head of Princeton's astronomy department and director of the Princeton University Observatory. He was something of a renaissance man – keen on music and an able mountaineer – but couldn't resist the occasional wild idea, such as climbing up the tower of the Princeton graduate college with ropes and pitons, only to be arrested by the university's security police. He cut something of an old-world figure with his upright bearing and courteous speech, but he was highly principled and always showed total independence of mind. Something that he could not resist, however, was a big and challenging scientific problem and controlled thermonuclear fusion fitted that bill perfectly.

In the tranquillity of Aspen, Spitzer had plenty of time to think. Like others before him, he realised that you would need a very hot plasma so that nuclei collide with enough force to fuse, that it would need to be relatively dense so that enough collisions take place, and that the best way of keeping it away from the walls of its container was with magnetic fields. Spitzer didn't, as Peter Thonemann and George Thomson had done, conclude that the pinch effect was the answer. Instead, while riding the long ski lifts up the mountain, he imagined a straight tube with magnetic field

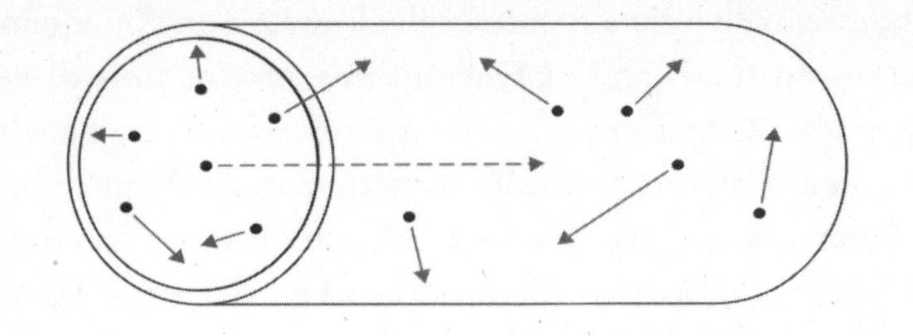

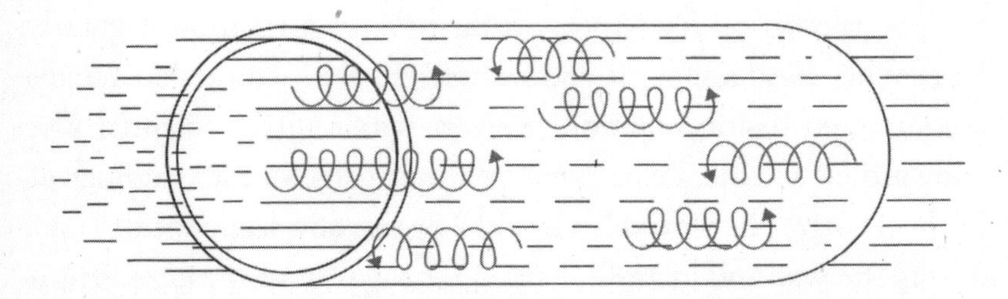

**Particles in plasmas, like in gases, naturally fly around in random directions. But apply a magnetic field and plasma particles become locked in spirals around the magnetic field lines.** (Courtesy of EFDA JET)

lines – imaginary lines that show the direction of the field at any place – running straight and parallel along its interior. The effect of this uniform field on the charge particles of the plasma – which would normally be moving in straight lines with random directions – would be to exert a force on each particle pulling it perpendicularly towards the magnetic field lines. This turns their straight line motions into tight little spirals orbiting around the field lines. The particles can move freely along the length of the tube, but are prevented from moving across it, towards the walls, by being bound to orbiting the field lines. As long as the field avoids the walls, Spitzer reasoned, so will the particles.

Creating a uniform magnetic field in a tube is easy: simply wind a length of wire round and round the tube along its full length and pass an electric current through it – a classic electromagnet. The problem is what to do with the ends? Because particles can move freely along the tube, they can just as freely drift out the end. Spitzer's solution was to make his tube endless by bending it around into a doughnut shape, or torus, just as Thonemann and Thomson proposed before him. This created a new problem, however: making the field bend around in a curve meant that it was no longer truly uniform. On the inside edge of the curve the turns of the electromagnet coil end up bunched closely together and on the outside edge they're spaced further apart. This means that the magnetic field the coils produce ends up stronger on the inside of the curve than on the outside. The effect on the particles of this non-uniform field is a vertical force which pushes the electrons up to the top of the tube and the ions down to the bottom, or vice versa. This separation of the electrons and ions creates an electric field which, in combination with the magnetic field, pushes the particles towards the outside edge of the curve. The overall effect is that the particles end up hitting the walls of the tube, the plasma loses energy and fusion doesn't happen. Spitzer calculated that particles would hit the wall before they had completed one circuit of the torus.

He was still puzzling over this conundrum when he left Aspen but it only took a few more days to come up with a solution: instead of a torus, build a figure-of-8 shape with two crossing straight sections joined by curved ends. With this arrangement, at one curved end the electrons are pushed up and the ions down and when they get to the other end the electrons are pushed down and ions up, so the drift of particles is essentially cancelled out.

Instead of getting down to work on the H-bomb with Project Matterhorn, Spitzer spent the next month writing up a detailed proposal for a power-producing thermonuclear reactor which he called a stellarator, and then sent it to the Atomic Energy Commission (AEC), the government agency set up in 1946 to manage both the development of nuclear energy and America's nuclear weapons, including Project Matterhorn.

Spitzer was invited to a meeting at the AEC in Washington on 11th May to discuss the issue of controlled fusion power. Also present was James Tuck who was at that time again working at Los Alamos on the H-bomb. Apart from Spitzer, there were probably few other minds in the US as primed and ready to tackle fusion power as Tuck's. Born and raised in Manchester, Tuck studied physical chemistry first at the Victoria University of Manchester and then at Oxford. Before finishing his doctoral thesis, he joined Leo Szilard working on particle accelerators at the Clarendon Laboratory. The war soon intervened: Szilard moved to the US and Tuck's head of department, Frederick Lindemann, was appointed to a senior post in the UK government, taking Tuck with him. Lindemann was already a personal friend of Winston Churchill and when the latter became prime minister the former was made his scientific adviser. Tuck, however, disliked the high politics of being a government science adviser, although he did make a valuable contribution to the war effort by developing shaped charges for use in armour-piercing shells.

Tuck's work on shaped charges got noticed in Los Alamos where they were having grave problems with the implosion mechanism of the plutonium bomb. Manhattan Project officials appealed, via Churchill, for Tuck to join their effort. He moved to New Mexico early in 1944 and helped to develop the explosive lens that was key to the bomb's success. He was present for the very first nuclear explosion at the Trinity test site in New Mexico in July 1945 and the first test at Bikini Atoll the following year. With the war over he had more time on his hands and so contacted Manchester University about finishing his doctoral thesis but was told that 'the statute of limitations on presenting a thesis had run out.' By this time his old boss Lindemann, now Lord Cherwell, was urging him to return to the UK. He arrived back at the Clarendon in autumn 1946, around the same time that Peter Thonemann arrived. Tuck helped install a new particle accelerator called a Betatron at the lab and, because of his fusion work at Los Alamos, was invited by John Cockcroft to the meeting at Harwell in January 1947 when George Thomson's pinch device was discussed.

So Tuck had a pretty good knowledge of all the fusion work that was going on in Britain when, in 1949, he got a call from Edward Teller to come to Los Alamos to work on the new H-bomb programme. Back in New Mexico, Tuck was set to work measuring the exact reactivity of the deuterium-deuterium and deuterium-tritium reactions but continued to think about constructing a pinch device along similar lines to Thonemann's. In spring 1951, a graduate student from Princeton spent some time at Los Alamos and told Tuck about Spitzer's plan for the stellarator. Tall and wire-thin, Tuck was a blunt northerner with an acerbic sense of humour, and one who wasn't shy of telling others when he didn't think their ideas were up to much. And that's exactly what he thought about Spitzer's stellarator.

For a start he believed that the stellarator, which aimed to hold its plasma at high temperature in a steady state for long periods, would lose too much heat by simple thermal conduction through the plasma and into the walls. The pinch, in contrast, was a pulsed device that squeezed the plasma very fast to achieve bursts of energy. He also distrusted Spitzer's optimism. He had seen firsthand some of the problems that Thonemann and Thomson had come up against and knew that achieving fusion would require a lot more than just devising a good confinement scheme. So many things were unknown or untested that it was far too soon to talk about power-producing reactors.

At the 11th May AEC meeting both the pinch and the stellarator were discussed. The evidence in support of both approaches was weighed up; both men interpreted it differently and both remained resolutely in favour of their own device. But the AEC officials liked Spitzer's plan and in July they awarded Spitzer $50,000 for theoretical studies into the stellarator as part of Project Matterhorn. Tuck was busy for most of the rest of 1951 with his studies of reaction rates but towards the end of the year he approached the head of the Los Alamos lab and proposed a new programme of controlled thermonuclear research there. He was awarded $50,000 of the lab's own funds to get things started. And so the seeds were sown for years of rivalry between Los Alamos and Princeton.

Meanwhile in Argentina, Perón grew tired of waiting to hear of more success from Huemul Island. He appointed a technical committee of physicists and engineers to investigate Richter's work. They reported to Perón in September 1952 that the temperatures Richter was achieving in his device were much too low to cause fusion reactions. Perón shut down the laboratory later that year. Whether or not Richter genuinely believed he could achieve fusion energy, he was working alone, cut off from the rest of the scientific community and so was unlikely to succeed. But this was not the last time in the history of fusion that extravagant claims turned out to be false.

Spitzer set to work refining the theory of the stellarator as well as sketching out a complete development plan. Work would start with a table-top device (Model A) which would show that a plasma could be created and confined and would heat the plasma's electrons to 1 million °C. The larger Model B would also heat the ions up to 1 million °C, while the Model C would be a virtual prototype power reactor able to reach thermonuclear temperatures of more than 100 million °C. The whole process would take about a decade.

In November 1951, the governing body of the AEC – five

commissioners appointed by the US president – considered whether controlled fusion should become a formal programme. In the optimistic spirit of the time, they were all generally in favour of pursuing new avenues of research. They were also aware that Britain was already working on fusion and suspected that the Soviet Union was too. After the shock of Russia's first atomic explosion two years earlier, the United States couldn't afford to fall behind in any important new field of research. It was a time of great insecurity for America: its national laboratories were racing to build the H-bomb before the Russians and its soldiers were fighting communist forces in the hills of Korea. The commissioners asked Thomas Johnson, director of the AEC's division of research, how much a fusion programme would cost. Plucking a figure out of the air he said that $1 million over three and a half to four years would prove whether fusion is feasible and if not, why not. So the programme was launched with a pot of $1 million.

Along with their generosity, the commissioners stipulated that the programme had to remain classified because of the potential of fusion neutrons for breeding plutonium, so the controlled fusion part of Project Matterhorn had to find a secluded place to do its work. Princeton University had recently acquired some property outside the city which had previously been owned by the Rockefeller Foundation. There Spitzer found a corrugated iron building that formerly housed laboratory animals. They would find out later that this 'rabbit hutch' was as hot as hell in the summer, but it would do. Windows were blackened, alarms fitted, barbed-wire fences erected and guards put on duty. Over the weeks that followed, Spitzer and his astrophysics colleague Martin Schwarzschild spent their weekends sitting on the floor of the hutch winding flat copper wire around 2-inch diameter glass tubes as they constructed Model A, America's first fusion reactor.

**Lyman Spitzer and his Model A stellarator.**
(Courtesy of Princeton Plasma Physics Laboratory)

Tuck didn't have to worry so much about security: he was working in a government lab surrounded by people who were already security screened. Especially useful to him would be all the expertise in measuring devices and instruments, or 'diagnostics,' that had been developed to analyse bomb tests. One of the major difficulties for fusion researchers at the time was finding out exactly what was happening inside a device. How dense and hot is the plasma, and is it humming along stably or wriggling out of control? Naturally, Tuck began working on a toroidal pinch. It had a pleasing simplicity to it which he hoped would mean a working reactor wouldn't be hugely expensive. Not everyone was convinced. A sceptical colleague referred to his device as an 'impossibilitron.' Tuck countered, 'Perhaps it will work and perhaps it won't,' and resolved to call it the Perhapsatron. Tuck scrounged some parts from a disused Betatron accelerator and got the lab's technicians to make a toroidal glass tube for the first Perhapsatron. His goals were straightforward: to see if he could produce a pinched plasma and if it is stable. If that worked he would aim to produce neutrons.

A third team also joined the effort. Edward Teller was agitating to set up a second weapons laboratory to work on the H-bomb in addition to Los Alamos. His chosen site was at Livermore, California, a small town in a cattle-rearing and wine-growing valley near San Francisco. The University of California at Berkeley bought an old military camp there in 1950 and built a lab. Herbert York, a young Berkeley physicist, was given the job of setting up the new weapons lab, so in the early months of 1952 York travelled to other labs in the H-bomb project, including Los Alamos and Princeton, to discuss Teller's plans. He was intrigued by the work of Spitzer and Tuck and decided that it would be a good idea if the Livermore lab had its own controlled fusion project, to broaden its programme beyond weapons work. York didn't want to simply duplicate the research elsewhere by building a pinch or a stellarator; he needed a device of his own. Spitzer's original idea, of a straight tube with an applied magnetic field running along the length of it, appealed to him because of its simplicity. The stellarator only got into problems because its tube was bent into curves to make it endless. What if you kept the simplicity of the straight tube and simply plugged the ends in some way to stop particles escaping? His first idea was to plug the ends with intense radio waves, which were known to exert some pressure on matter.

York gave a series of lectures at Berkeley's Radiation Laboratory on the problems of fusion. In the audience was Richard Post, a recent recruit to the lab. Post was fascinated and loved the fact that it was an area of research with a socially useful goal. His recent doctoral research had involved both radio waves and plasma so he immediately wrote a long memo to York analysing the radio wave plug idea and other possibilities. York called him in and offered him the job of heading up the new lab's controlled fusion group. Post and some colleagues set to work on the theory of such a device. He soon concluded that it would take a huge amount of power to produce radio waves intense enough to plug the ends of the tube. They began looking at the magnetic fields themselves and found that if the field was strengthened at the ends of the tube – with extra windings of the electromagnet – this forced the field lines closer together. If the conditions of the plasma were right, this crimping of the field at the ends forced the moving particles to reverse direction, an effect known as a magnetic mirror. York and Post now had a fusion device they could call their own.

During the course of 1952, both the stellarator and the Perhapsatron were built and powered up for the first time. Both teams started off filling their tubes with noble gases rather than deuterium because at first they weren't trying to cause fusion reactions but just to show that they could create a plasma, confine it and heat it. Spitzer was able to produce a plasma in his Model A and, applying a stronger magnetic field, was able to show that the confinement was improved. But particles were still drifting to the wall faster than theorists had predicted, so Spitzer still had a puzzle on his hands. In their theoretical models, they had assumed that all the particles in the plasma moved independently of each other, influenced predominantly by just the magnetic field. That assumption was something of a necessity because to accept that the particles influenced each other as well would make their calculations impossibly complicated in an age before computers were widely available. If the particles do influence each other then 'collective effects,' such as waves and turbulence, would have to be accounted for.

The Perhapsatron was also proving to be more of a conundrum than Tuck had expected. When the device, which was about a metre across, was powered up the researchers could see a glowing plasma through windows in the side and when they induced a current in the plasma they could see for a brief instant the plasma being pinched into a fine filament at the centre of the tube before it suddenly disappeared. Using high-speed cameras that take shots at various intervals after the pinch was applied, they could see that it remained stable for just a few millionths of a second before the filament began to violently thrash about and break up. Tuck had come up against the kink instability, just as Bob Carruthers did at Harwell. Tuck grasped the fact that there was a brief period of stability before the plasma broke up and concluded that he could either focus on making the pinches very fast – getting a burst of energy before the instability kicks in – or on finding some way to stabilise the plasma.

With Livermore's magnetic mirror work also progressing, the fusion programme was gaining some momentum. It had its first conference at the end of June 1952 in Denver, with eighty attendees. It also acquired a name: Johnson at the AEC was plotting to close down another research programme called Project Lincoln at the Hood Laboratory and give the money to Tuck. With that conjunction of names, it just had to be Project Sherwood. But it remained a pretty relaxed affair. The priorities of the programme were set by the researchers themselves. There were only a few dozen people in total working on fusion and many of them, including Spitzer, Tuck and Post, were only working on it part-time; they had other lines of research going on in parallel. Nevertheless, none of them ever doubted, despite the severe lack of knowledge of the intricacies of plasma behaviour, that it would be a simple progression over the course of a decade or so from table-top demonstrator devices to larger experimental reactors and then prototype power plants.

* * *

The appointment of Admiral Lewis L. Strauss (pronounced 'straws') as chairman of the AEC in mid 1953 stirred up the cosy club of fusion research. Strauss was a man of his time: a firm believer in technology and US scientific pre-eminence, and intensely suspicious of the Soviet Union. He wanted fusion to be achieved to show the power of the capitalist system, and he wanted it done during his tenure at the AEC.

Strauss was born and raised in Virginia and, on graduation from high school, intended to study physics at the University of Virginia. But problems with his family's shoe wholesale business meant that he had to go and work with his father as a salesman. Although after three years he had earned enough to go to college, in 1917 he volunteered to work with Republican politician Herbert Hoover (a future US president) to support the war in Europe. He made important political contacts while working as Hoover's private secretary and after the war worked to help Jewish refugees from Europe, an experience that gave him a profound dislike of communism. He never made it to college to study physics but instead joined a New York banking firm, becoming a partner in 1929 and making his fortune. A long-time member of the Navy Reserve, he volunteered for active duty in World War II and managed naval munitions, eventually reaching the rank of rear admiral.

Strauss maintained an interest in physics and, following the death of his parents from cancer in the 1930s, set up a fund to support research into radiation treatment of the disease. Through the fund he met prominent physicists of the time, including Leo Szilard. Strauss funded some work by Szilard into fission chain reactions, which would eventually lead to the scientist's warning to President Roosevelt in 1939 about the possibility of an atomic bomb.

When President Truman formed the AEC in 1947, Strauss was an obvious choice to be one of the first commissioners.

As a commissioner, Strauss came into contact with J. Robert Oppenheimer, former scientific leader of the Manhattan Project. Oppenheimer was a totemic figure among many postwar scientists. Not only had he successfully led the scientific effort to design the atomic bombs that hastened the end of the war, but afterwards he had campaigned against a nuclear arms race and argued for the international control of nuclear materials and technology. Oppenheimer was the chair of the AEC's General Advisory Committee and his liberal views were like a red flag to the conservative Strauss. Strauss was a strong advocate of moving straight on to building an H-bomb, as an ultimate deterrent to Soviet aggression. Oppenheimer, in the 1940s, didn't think it would work and instead favoured a stockpile of smaller tactical atomic bombs. Strauss was alarmed by the unmasking of Soviet spies linked to the Manhattan Project, including Klaus Fuchs and Julius and Ethel Rosenberg. Oppenheimer was known to have had friends, relatives and colleagues who were communists during the 1930s and many, including Strauss, believed he was an unacceptable security risk. During a congressional hearing at which Strauss lobbied against allowing the export of radioactive isotopes, Oppenheimer mocked him by stating that the isotopes were 'less important than electronic devices but more important than, let us say, vitamins.'

Once the crash programme to develop the H-bomb was approved by President Truman in 1950 Strauss stepped down from the AEC but he did not forget his animosity towards Oppenheimer. One of Strauss' first acts when he returned to the AEC as chair in 1953 was to ask the FBI to intensify its surveillance of Oppenheimer, which included following his movements, bugs in his home and office, telephone taps and opening his correspondence. The FBI didn't find any evidence of disloyalty but Strauss persisted. In December he told Oppenheimer that his security clearance had been revoked and showed him a list of charges against him. Oppenheimer refused to resign and so during the following April and May hearings were held into his security clearance. All of this was taking place during the anti-communist witch-hunts of Senator Joe McCarthy and hearings to investigate such a high-profile figure as Oppenheimer drew huge publicity.

The hearings raked over Oppenheimer's prewar communist ties and his association with Soviet spies who had worked at Los Alamos. Many prominent scientists, politicians and military officials spoke in his defence. Teller testified, saying that he thought Oppenheimer was loyal to the United States but that his judgment was so questionable that he should be stripped of clearance. Oppenheimer's own testimony was often inconsistent and erratic and in the end his clearance was revoked. He was widely seen as a victim of anti-communist hysteria; Teller, on account of his testimony, was shunned by the scientific community; and Strauss won a reputation as a McCarthyite witch-hunter.

Back at the AEC, Strauss was unhappy about the slow progress on Project Sherwood. He called a meeting of the project leaders from Princeton, Los Alamos and Livermore plus a few other prominent physicists to discuss the future of the programme. This group of advisers proved to be more cautious than Strauss liked. They described Sherwood as a long-term effort and felt it was too early to speculate about large-scale reactors. They recommended continuing on the same course as before. Strauss had other ideas. He told Johnson to beef up the organisation of Project Sherwood within AEC headquarters and prepare to treble the project's spending. In addition, despite President Eisenhower's Atoms for Peace speech in December and a general move towards nuclear declassification, Strauss insisted on tight security.

Johnson appointed one of his staff, Amasa Bishop, to work on managing the project full-time. He also set up a steering committee – made up of the project leaders from each laboratory plus Edward Teller – which met several times a year. With the conferences of Project Sherwood happening at a similar frequency, there was suddenly a sense of urgency among the researchers to get results ready for the next meeting. The numbers working on Sherwood projects grew quickly, from forty-five in March 1954 to 110 a year later and double that by 1956. Strauss lavished money on the project, so much so that researchers almost didn't know how to spend it all. From the $1 million that Johnson originally secured for 1951-53, the Sherwood budget increased to $1.7 million for 1954 alone, $4.7 million in 1955, $6.7 million in 1956 and $10.7 million in 1957. Strauss even joked about offering a $1 million prize to the first person or group to achieve controlled thermonuclear fusion.

Johnson decided that the stellarator was the most promising of the different approaches and it should be made the focus of Project Sherwood. Model A was showing good confinement and Model B was under construction. This second machine would have a much stronger magnetic field than its predecessor and, Spitzer hoped, would reach ion temperatures of 1 million °C. But in the new regime instituted by Strauss, Spitzer and the other project leaders were starting to feel pressure from the head office. Johnson wanted the Princeton team to start working on additional heating systems for Model B to help it reach higher temperatures. Spitzer, however, was reluctant to do this until they had got Model B running and could see how it performed. Johnson insisted that there was no time to waste. He also wanted the Princeton team to get started on a design for Model C, Spitzer's putative prototype power reactor.

From Spitzer's point of view, this was becoming less like a research programme and more like a crash development effort – like the one then going on to develop the H-bomb – where instead of progressing methodically and in sequence, things are done in parallel to save time. But there are dangers in this approach. Moving onto the next thing before Model B is shown to work runs the risk that they may go down the wrong path and have to retrace their steps. Despite Spitzer's reservations, the Princeton team started work on a design for Model C in the autumn of 1954, a matter of weeks after Model B began operating.

It was the pinch devices that were soon getting the attention, however. A young theorist called Marshall Rosenbluth had arrived at Los Alamos. He had studied for his PhD at the University of Chicago tutored by Edward Teller, who later recruited him to join the H-bomb team. He was not entirely happy about doing bomb work but when he found out what Tuck was doing at Los Alamos he was keen to take part. He and others began working on a more rigorous theoretical model of how a pinched plasma works and came up with M (for motor) theory. This had a dramatic effect. Suddenly the researchers had a better understanding of what the plasma was doing. One of M-theory's first predictions was that if the plasma current was made stronger, the pinch will heat the plasma to a higher temperature. In a toroidal pinch device like the Perhapsatron, that would require the electromagnet to create a stronger magnetic field, which was tricky at the time. So Tuck and the Los Alamos researchers began building a straight pinch device in which it would be possible to create a strong current by putting an electrode at each end and a voltage across them. Another team at the University of California, Berkeley, did the same.

By the autumn of 1955, Tuck had a 1m-long pinch device called Columbus along which he could apply 100,000 volts. Much to their surprise, every time the researchers produced a pulse of current to create a pinch they detected millions of neutrons coming out of the plasma. If those neutrons were being produced by thermonuclear reactions, the plasma must be reaching temperatures of millions or tens of millions of °C. There was much excitement throughout the community of Sherwood scientists about the pinch's success. Tuck, who had kept his goals modest, now started to form grander plans. A power-producing reactor based on a straight pinch would still have to operate in fast pulses, producing a burst of fusion energy before instabilities ripped the plasma apart. He did some rough calculations and estimated that using a large-bore tube an electric pulse equivalent in energy to a ton of TNT would produce several tonnes equivalent of fusion energy.

Not everyone was convinced, however, that the neutrons produced in the straight pinches were from thermonuclear reactions. At a Sherwood conference in October 1955, Stirling Colgate, a researcher from Livermore, pointed out that the pinches were sometimes producing neutrons in conditions which M-theory predicted would create temperatures too low for fusion. Colgate suggested that the pinch teams should try measuring whether the same number of neutrons came out in all directions, as should be the case if the source is a thermonuclear plasma – the same test that ZETA would later be subjected to. The pinch team at Berkeley did the test and found that many more neutrons emerged from one end of the tube than the other. So the neutrons were not thermonuclear: some ions in the plasma were being accelerated to high speed in one direction, producing neutrons by fusion but leaving the bulk of the plasma at too low a temperature. The following spring, when Kurchatov came to Harwell and gave his unexpected lecture about thermonuclear fusion, western scientists learned that their Soviet colleagues had experienced the same surge of hope followed by disappointment when they had detected neutrons from one of their devices in 1952.

This did not kill off US interest in the pinch, however. A couple of ideas for stabilising a pinched plasma had been kicking around for a few years – ideas that Peter Thonemann and his colleagues at Harwell had also incorporated into ZETA. The first was the idea of making the vessel from a conducting metal, or enclosing it in a conducting shell, to repel particles from the walls. The second was the idea of applying a longitudinal magnetic field along the tube to give the plasma a stabilising magnetic 'backbone.'

Rosenbluth, who would eventually earn himself the nickname 'the pope of plasma physics,' incorporated these modifications into his M-theory model of pinches and predicted that, under certain narrow conditions, a pinched plasma could be made stable. By the summer of 1956, Tuck's team at Los Alamos and the Berkeley group were both showing improved stability using Rosenbluth's prescription. This opened up exciting possibilities: maybe a pinch didn't have to be a fast-pulsed device to get some fusion before instability kicked in; maybe it could hold onto its plasma for longer.

Strauss' effort to get faster results from Project Sherwood seemed to be working, but cracks were starting to appear in the cocoon of secrecy he had wrapped around the project. The Atomic Energy Act of 1954 allowed the US to again share nuclear secrets with allies such as Britain. In line with President Eisenhower's Atoms for Peace policy, an increasing number of formerly secret nuclear projects were edging out into the light of day. A huge amount was revealed at the first Atoms for Peace conference in Geneva in August 1955 and Homi Bhabha's prediction about fusion energy put governments under pressure to reveal more. Not wanting to leave the United States looking like it was lagging behind other nations, Strauss was forced to announce the existence of Project Sherwood the next day, but no other details were revealed.

When plans were announced for a follow-up conference in Geneva in 1958, the US administration decided that the AEC should put up a show-stopping exhibit to demonstrate the preeminence of American nuclear technology. At first, fusion hadn't been considered seriously as a candidate for the 'spectacular,' but the cost estimates that the AEC commissioners got for displaying other types of reactor proved sky high. Something from Project Sherwood only looked possible because declassification of the programme was starting to seem inevitable. The military justification for maintaining secrecy – that a fusion neutron source could be used to make plutonium for bombs – was evaporating as mining companies found ever larger natural reserves of uranium.

Johnson and Bishop in the AEC's research division had long been pressing for declassification. In April 1956 they had recommended that fusion information should start to be exchanged with the British and the following month recommended full declassification. Most of the AEC commissioners supported the proposal, but Strauss did not. In September they suggested a slightly watered down proposal: declassify everything apart from any details crucial to a working fusion reactor. To get the most out of the announcement, it should be made just before a major fusion conference to encourage the Soviets to make similar revelations. Strauss still opposed the suggestion, but the commissioners unusually decided to go ahead on a majority vote. By the following month, the first delegation from Harwell arrived to tour the main fusion labs. The US team that visited the UK in November was clearly wowed by ZETA. This was no proof-of-principle lab experiment; ZETA looked like something that could generate power. The size and ambition of the machine was far ahead of any of the American projects and the team members were concerned that, if Britain decided to ship ZETA to Geneva as an exhibit, any Sherwood effort would be overshadowed. Beyond the rivalry between Princeton and Los Alamos, the US teams now had the British and, following Kurchatov's lecture, the Russians to worry about.

US unease about ZETA increased when it began operating in August 1957 and almost immediately started producing neutrons. Although news of the breakthrough did leak to the press in early September, Strauss was determined to keep a lid on it. He was convinced that the US programme was superior and, although the work on some devices had been slowed by technical problems, both the straight and toroidal pinches were showing promise. The declassification plan suggested that nothing be revealed until the 1958 Geneva conference – a whole year away. If Strauss could induce the British to keep quiet about ZETA until then, Sherwood would have time to catch up and the US exhibit would be the star of Geneva.

But the project leaders at the labs and the AEC commissioners were in two minds about a spectacular fusion exhibit at Geneva. It would bring in extra funding from the government but the months spent preparing the exhibits would distract the scientists from the real business of mastering fusion. And there were still serious doubts that they could achieve thermonuclear neutrons in time. The situation was complicated further by the shock arrival of Sputnik on 5th October. Now Strauss had the added pressure of wanting to make some public demonstration of US technical superiority to trump the Soviet feat. A crisis meeting was called for 19th October to make a decision about Geneva. The project leaders as well as Johnson and Bishop from the AEC had decided in advance that aiming to show a reactor producing thermonuclear neutrons was just too risky and some other lower profile exhibits should be planned. Most of the commissioners didn't like the odds either. But Strauss was adamant. He wanted fusion to be achieved on his watch, he wanted the US exhibit to overwhelm all others at Geneva, and he wanted to do his bit to restore American pride. So thermonuclear neutrons would be the centrepiece at Geneva. This was a turning point for the US fusion programme: scientists were no longer setting its goals or pace; it had become an instrument of US foreign policy.

While Strauss was able to assert himself over the Geneva exhibit, he was not getting his way over ZETA. The British press had got hold of the fact that the ZETA team was being kept quiet because of US insistence on secrecy and they were painting Strauss as the villain. With newspapers clamouring, the Harwell scientists didn't know how long they could keep a lid on ZETA. The British wanted to publish their results, and they wanted to do it before Geneva. Strauss finally gave in to the inevitable. To try and lessen the impression that the UK was ahead in the fusion race he resolved to publish details of the US pinch experiments at the same time. But first he sent the team of American researchers to take a look at ZETA to find out if it really was producing thermonuclear neutrons.

The Americans were, by and large, convinced of ZETA's neutrons. Shortly afterwards the latest versions of the Perhapsatron and the Columbus straight pinch started producing neutrons too. Altogether, things were looking very bright for the pinches. Some had their doubts, however. Spitzer was sceptical that ZETA was really reaching temperatures as high as 5 million °C during such a short pulse. He suspected that a small volume of plasma was being heated more than the rest by an instability. Colgate at Livermore was also concerned about the British temperature claims. He, along with a couple of colleagues, made a careful comparison of all the pinches that were then producing neutrons, including ZETA and Britain's Sceptre-III at AEI, the Perhapsatron and Columbus. Instead of focusing on the neutrons they looked at the electrical conductivity of the plasma in each case. Using a formula linking conductivity and temperature they concluded that the temperature in ZETA was probably a tenth of Harwell's claim, so its neutrons could not be thermonuclear.

Nevertheless, Strauss and the US fusion scientists could only grit their teeth in late January when the results from ZETA and the American pinch experiments were published together. As Strauss had feared, the US results were largely ignored and the Harwell scientists were hailed as the conquerors of fusion.

ZETA's fall from grace a few months later shocked US scientists, even though some had doubted its results. The team at Los Alamos soon carried out similar tests on the Perhapsatron and Columbus to the ones that sank ZETA. Their machines too were shown to be producing spurious neutrons. That left Project Sherwood without a triumphant centrepiece for the Geneva conference. But Strauss was undaunted. He wanted neutron-producing devices whether the neutrons were thermonuclear or not. Strauss expected that no one at the meeting would be in a position to tell the difference and if the Soviets had a device spitting out neutrons and the US didn't it would be a disaster. So he kept up the pressure on the American teams to get their exhibits ready.

Over the summer, dozens of researchers in all the fusion labs laboured over models and displays and tested working devices before boxing them up and despatching them to Geneva. In all, some 450 tonnes of equipment was shipped from the four Sherwood labs. (Oak Ridge National Laboratory's previously small fusion research effort grew rapidly in the run-up to Geneva and it became a fully-fledged member of the project.) Whole airliners were chartered to carry Sherwood technicians and their families to Switzerland, some arriving more than a month early to get everything set up before the 1st September start of the meeting. The number of people registered to attend dwarfed the 6,500 hotel beds in Geneva, so some participants had to stay in cities such as Evian, 50 kilometres away.

The scale of the conference was huge. It was held in the Palais des Nations, built in the 1930s for the League of Nations but later home to the United Nations' European headquarters. Its main assembly hall could hold around 2,000. In the grounds of the palace an enormous exhibition hall was specially built for the conference. Five thousand scientists from sixty-seven countries attended the meeting, along with 900 journalists and 3,600 observers from industry. The interested public could also look around the exhibition. *Time* magazine called it the 'monster conference.'

While the US exhibit lacked the knockout display of fusion neutrons that Strauss had wanted, it was certainly impressive. Princeton had shipped over its Model B-2 stellarator and displayed mock-ups of the Model C and other devices; Oak Ridge had a model of its Direct Current Experiment (DCX) device; models of a mirror machine and linear pinches were on show from Livermore and Berkeley; and Los Alamos had a working Perhapsatron and a Columbus both producing neutrons, as was a variant of the pinch known as Scylla. Along with displays of fission reactors and other nuclear technologies, the US exhibit took up more than half of the exhibition space and registered 100,000 visitors. It had cost $4.5 million – a huge sum in 1958.

Two days before the conference opened both the US and the UK had announced total declassification of their fusion programmes. The scores of fusion researchers who had been working in secret laboratories in those countries and the Soviet Union emerged blinking into the light. They could present papers to a wide audience in the assembly hall and also had to deal with press conferences and curious members of the public. At a time when East and West were locked together in a Cold War, at Geneva you could see Russian and American scientists, heads together, earnestly discussing plasma physics in the conference corridors, or sitting on the grass in the palace gardens among the peacocks that lived there.

Despite the excitement generated by this entire new field of science suddenly revealed, the scientific presentations in the assembly hall had a more reflective tone. The disappointments earlier in the year with the pinch devices – ZETA, Perhapsatron and Columbus – along with the technical difficulties being experienced by other machines, had caused many to rethink their optimistic forecasts of a smooth progression from proof-of-principle to prototype to power reactor. Peter Thonemann told the assembled scientists:

> To my mind, the problem of stability is of paramount importance. Unless the rate at which charged particles cross mag-

> netic lines of force can be reduced to that given by classical diffusion theory, the loss of energy to the walls will prevent fusion reactions from becoming a practical power source.
>
> I think that the papers to be presented at this Conference, and the discussions which follow them, will show that it is still impossible to answer the question, 'Can electrical power be generated using the light elements as fuel by themselves?'

Edward Teller listed a number of hurdles that would have to be overcome, including the intense neutron radiation emanating from fusion reactions that changes the nature of the materials of the reactor and makes the reactor a no-go area for human operators.

> These and other difficulties are likely to make the released energy so costly that an economic exploitation of controlled thermonuclear reactions may not turn out to be possible before the end of the 20th century.

After two weeks of discussions, few scientists will have returned home from Geneva with any illusions about the difficult tasks facing them.

The Geneva conference marked something of a watershed for the US fusion programme. In June, just before the meeting, Lewis Strauss had stood down as head of the AEC, although he still led the delegation to the conference. He had not achieved his goal of seeing a successful demonstration of fusion power during his tenure, but he had taken a somewhat relaxed small-scale research programme involving a few dozen people (fiscal year 1954 budget: $2 million) and pumped it up into a dynamic development effort employing hundreds (1958 budget: $29 million).

The atmosphere of suspicion and competition between the US, UK and Soviet programmes, fuelled by the secrecy imposed on them, was diffused when the researchers from each side actually met, compared notes, and realised that they were all in the same scientific boat: desperately seeking a route to fusion energy while not really having a good enough grasp of how plasma behaves. The relationships forged at Geneva meant that, even while most scientific disciplines and most of normal life were divided into East and West, fusion research remained a truly international activity. Scientists were able to visit each others' labs across the Iron Curtain relatively easily and international conferences brought them all together on a regular basis.

East-West links were far from the only changes to come from declassification. Overnight the researchers' top-secret project became a normal scientific discipline. Princeton's Project Matterhorn was renamed the Princeton Plasma Physics Laboratory. It became easier to hire new staff and to train new researchers. Princeton and the Massachusetts Institute of Technology set up graduate programmes in fusion research.

Fusion was absorbed into mainstream science. Researchers in closely related fields could now make inputs and attend fusion conferences, which broadened the available expertise. The American Physical Society set up a Division of Plasma Physics which held its own meetings. Specialist plasma physics journals were created and the papers submitted were subjected to 'peer review' so that they were vetted by other scientists. The effect of all these changes was to raise the scientific standards of fusion research; there emerged new levels of scrutiny and criticism that, if they had arrived earlier, might have tempered the unbridled optimism that surrounded ZETA, Columbus and the Perhapsatron.

Industry also got in on the act. Westinghouse, which had seconded two engineers to work on the design of Princeton's Model C, let them remain there while keeping them on its own payroll.

Thinking that fusion could one day rival fission energy, General Electric and the recently formed General Atomics both started up their own fusion research programmes.

Project Sherwood also had to face the new reality of its position within the AEC. Under Strauss' tutelage, it had led a charmed existence with ever increasing budgets, because he had a special interest in fusion. The new AEC chairman, John McCone, had no such soft spot. In fact, for his first year in the post his attention was focused elsewhere, on fission reactors, nuclear-powered ships and his concern that a ban on nuclear testing might be imposed. But in July 1959 he ordered a review of Project Sherwood, in particular of its level of funding. Sherwood's budget had been tripled in November 1957 to help the labs prepare for the Geneva conference, and the lab heads were led to believe that this level of funding would continue after the meeting.

McCone had other ideas. The project should be funded, he felt, in line with its value to the AEC, just as any of its other research programmes were. He also could not understand why there were so many different fusion devices being studied. Why not just focus on the few most promising ones? The urgency of the project had also abated. While once it had looked a couple of steps away from a power-producing reactor, now it seemed more focused on the study of fundamental plasma physics. The Sherwood steering committee argued successfully for retaining all the different device types, but it did accept that a 10% cut in funding was inevitable.

Pinches fell from favour in the United States following the disappointments of Columbus and the Perhapsatron and hopes were pinned on the stellarator. Princeton's Model B was produced in various versions (B-2, B-3) to correct magnet faults and add extra heating systems but all of them suffered from a problem that the scientists called 'pumpout.' Essentially, particles were drifting to

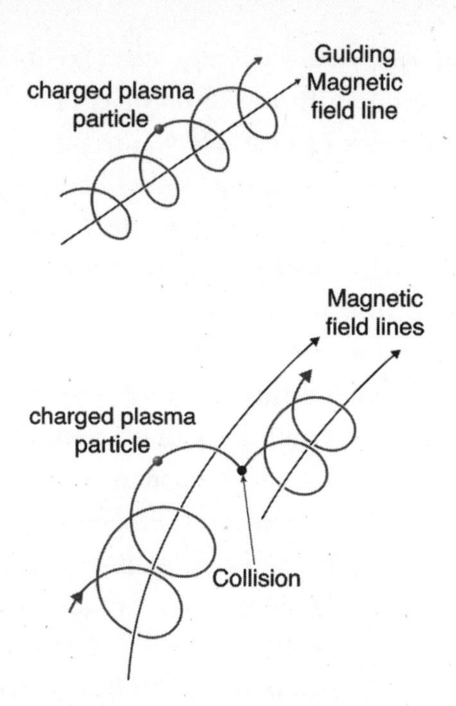


**Spiralling plasma particles can jump to another field line following a collision, and so diffuse across the field towards the vessel wall.** (Courtesy of EFDA JET)

the edge of the containment vessel faster than predicted by theory and taking heat out of the plasma with them.

In the simplest picture of plasma in a stellarator, particles are held in place by the magnetic fields, making tight spirals around the magnetic field lines. But this doesn't take into account collisions between particles. When this happens particles can be bumped from one field line to another and so gradually drift across the lines. This process – diffusion – happens in all gases, but fusion researchers had counted on magnetic fields slowing it down. Theorists predicted that the rate of diffusion would decrease in line with the square of the magnetic field strength. So doubling the field strength cuts the diffusion rate by four. This situation, known as classical diffusion, was not what they saw in practice.

During the war, American physicist David Bohm had done some research on the diffusion of plasmas in magnetic fields as part of the Manhattan Project. He wanted to separate isotopes of uranium so that natural uranium could be enriched for use in atom bombs. Purely by experiment he found that diffusion across magnetic fields decreased in proportion with the field strength, not field strength squared. This relation was much less favourable for fusion reactors because increasing the field strength does not rein in diffusion so strongly. Builders of fusion machines in the 1950s didn't think that Bohm's formula applied to them because he had studied plasmas of much heavier ions, at lower densities and temperatures. Pumpout was causing them to have doubts. The team working with Model B-3 studied the rate of diffusion over a wide variety of magnetic field strengths and found they varied exactly in the way Bohm had seen.

In May 1961, after four and a half years of design and construction, Princeton completed the Model C stellarator. Its race-track-shaped plasma vessel was 12m in length. Because of its larger size, the Model C held onto particles ten times longer than the Model B-3 had, but when researchers investigated the effect of increasing magnetic field strength, Bohm's relation held sway rather than classical diffusion. Throughout the late 1950s and early 1960s, Bohm diffusion was the bane of fusion scientists' lives. All they could do was go back to the drawing board, try to understand plasma better and see if they could find a way around the problem.

It was a bad time for the fusion programme to show signs of doubt. The shock of Sputnik's launch in 1957 had prompted a huge hike in government spending on science and technology, but that was now being called into question. More specifically, it was more than a decade since the AEC started funding controlled fusion

**The end of the line: the Model C stellarator.**
(Courtesy of Princeton Plasma Physics Laboratory)

research and scientists had promised prototype power reactors by this time. Instead, money that was meant to be spent on developing a new source of energy was paying for basic research on plasma physics. In 1963 the US Congress trimmed $300,000 from the programme's $24.2 million budget. The next year it was cut again to $21 million. Even within the AEC fusion had few supporters. The commission was more concerned with developing a fast breeder reactor, which seemed a better short-term prospect for energy production than fusion.

In September 1965 fusion researchers from across the world gathered at the Culham Laboratory, the new home of

Britain's fusion programme, for a conference organised by the International Atomic Energy Agency. The IAEA had started the meetings to sustain the international cooperation begun at Geneva in 1958. The first had been at Salzburg in 1961 and now it was Culham's turn. Spitzer gave a presentation summarising the results from all the toroidal machines worldwide. It was not an encouraging survey: all seemed to be failing either because of Bohm diffusion (the stellarators) or instabilities (the pinches). There was one potential bright spot piercing the gloom of Spitzer's talk: Lev Artsimovich, the forthright and combative head of the Soviet fusion effort, described encouraging results from Russia's variant on the toroidal pinch, known as a tokamak.

The tokamak takes its name from the Russian phrase *toroidal'naya kamera s magnitnymi katushkami* (toroidal chamber with magnetic coils). It has a plasma current driven around the torus by an electromagnet and so its plasma is pinched into a narrow band around the centre of the tube. Where it differs from other pinch devices is in the added longitudinal magnetic field directed around the ring. Such fields were added to the Perhapsatron and ZETA to help stabilise the plasma, but the tokamak's longitudinal field was around 500 times stronger. This field, combined with the one produced by the plasma current, resulted in a field which twists in a helix as it moves around the ring – a magnetic corset holding the plasma in place. Artsimovich told his fellow scientists at Culham that the tokamak was proving capable of suppressing instabilities and that it was confining the plasma ten times longer than Bohm's formula said was possible.

The Russians' Achilles heel was their measurements, however. Their instruments were primitive and they couldn't take readings of temperature and confinement time directly; instead they inferred them from other measurements. As a result, many at the meeting were sceptical of the seemingly huge advances that the

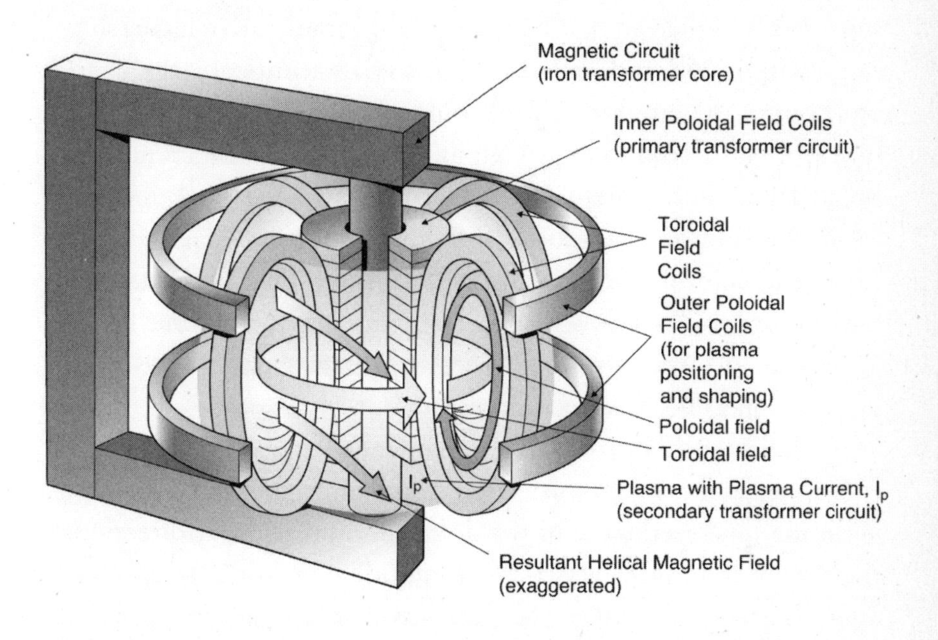


**In a tokamak, the horizontal toroidal field and vertical poloidal field combine to produce helical magnetic field lines.**

(Courtesy of EFDA JET)

Russians had achieved with the tokamak, sparking an acrimonious debate between Spitzer and Artsimovich. Spitzer said he felt deep pessimism for the future and many delegates wondered if instabilities and Bohm diffusion were inescapable properties of hot plasma.

Back home in the United States, Spitzer was soon forced to make a difficult personal decision. While he had worked on fusion, he had retained his post as head of Princeton University's astronomy department. But now the university asked him to become chair of its general research board. Three jobs seemed like too much. When Spitzer had devised the stellarator on the ski lifts of Aspen, he considered it a ten-year project to get to a prototype power reactor. Now, fifteen years on, the stellarator still had serious problems and prototypes were nowhere to be seen. Although some years earlier he had handed over administrative leadership of the lab to Melvin Gottlieb, he was still scientifically the heart and soul of the laboratory and could often be seen riding on his bike from Princeton to the lab along the historic Route 1 highway. So, in 1966, Spitzer stepped down from his post at the Princeton fusion lab and returned to his first love: astronomy. He went on to have a major impact in that field, devising some of the earliest orbiting space observatories and helping to get the Hubble Space Telescope off the ground. NASA's Spitzer Space Telescope, launched in 2003, is named in his honour.

However, his former fusion colleagues now had to face a thorough review of the fusion programme ordered by the AEC. With the field seemingly in the doldrums and without direction, the review committee was asked to decide if it should be accelerated, decelerated or killed. Members visited labs, heard presentations and debated the relative merits of different machines. The committee concluded that fusion research was worth preserving. It acknowledged that progress had been made and argued that if fusion proved possible, the US needed a cadre of qualified scientists to take advantage of it. It also argued that to lose the race for fusion to some other nation was on a par with losing the space race to the Russians. The panellists recommended that the budget for fusion be increased by 15% per year for five years, but the prospect of Congress agreeing didn't look good.

The United States at the time was increasingly committed to the war in Vietnam and President Lyndon B. Johnson was trying to push through a set of domestic reforms known as the Great Society which aimed to tackle poverty and racial injustice with increased spending on education, medical care and urban problems. After much debate within the AEC, in Congress and in the White House, the AEC commissioners came up with a more modest proposal: the budget would rise by 15% in the first year followed by progressively smaller rises until it was boosted by around 6% in the fifth year. In addition, up to $4 million was to be budgeted each year for equipment.

With the stellarator struggling, researchers were still trying to find other more successful confinement schemes. Livermore and General Atomics were still working on mirror machines. At Los Alamos, Tuck and his team were developing a variation on the toroidal pinch called Scyllac. And General Atomics had also teamed up with the University of Wisconsin on an elaborate device called a multipole that showed good confinement but only at low density and temperature.

But when researchers gathered at Novosibirsk in Siberia for the third IAEA fusion conference in August 1968, all the talk was about tokamaks. While the results that Lev Artsimovich had reported in 1965 at Culham had been impressive, the latest readings from the T-3 and TM-3 tokamaks were sensational. Again, the temperatures were not measured directly but the Russians calculated them to be more than 10 million °C, ten times that of any other toroidal machine. The confinement showed a similar improvement. Artsimovich was rigorously grilled by sceptical US researchers, particularly those from Princeton. They argued that the Russians were probably measuring the temperature of a small group of electrons that had been accelerated to a high speed, and were separate from the rest of the plasma – the same problem that had led to false optimism about ZETA, Columbus and the Perhapsatron.

Throughout the 1960s, the Russians had developed a close collaboration with the British fusion effort at Culham. The British, then led by Sebastian Pease, were still working with ZETA, a pinch device, and the Russian tokamaks were also a variant of the pinch, so they had things in common. Meanwhile, the Culham researchers had also perfected a new technique to directly measure the temperature of a hot plasma. They would shine a bright laser beam into the plasma and by analysing the spectrum of light that is bounced out by the particles they could accurately calculate the plasma temperature. At Novosibirsk, Artsimovich made a bold proposal to Pease: send a British team to Moscow with their laser thermometer, measure the temperature in the T-3 tokamak and settle the issue once and for all.

Pease travelled home and won the agreement of the British government to send his staff, and technology, to the Soviet Union. At the height of the Cold War, this unprecedented East-West collaboration would decide whether the tokamak was another embarrassment, along the lines of ZETA or Juan Peron's Thermotron, or whether it was the breakthrough fusion had been searching for.

CHAPTER 4

# Russia: Artsimovich and the Tokamak

WIND BACK THE CLOCK AGAIN, THIS TIME TO 1949 AND Sakhalin, a cold, remote and inhospitable island off Russia's far eastern coast. Soviet forces captured a large part of the island from the Japanese in the closing days of World War II and for the soldiers stationed there after the war it must have been a desolate posting. But to Red Army sergeant Oleg Aleksandrovich Lavrentyev it offered what he needed: time to study. Lavrentyev had not finished school when he joined the Red Army at the age of 18, but what he had read about the fission of uranium and the possibility of a nuclear chain reaction excited him and he was determined to study physics. A veteran of battles against the Germans in the Baltic States during 1944 and 1945, he was posted to the east after the war. Working as a radiotelegraph operator, he had plenty of time to read. He got hold of physics textbooks, monographs and even subscribed to the academic journal *Soviet Physics-Uspekhi*. He would give reports and lectures to his officers on scientific and technical subjects. In 1949 he completed his school exams, having covered three years' study in twelve months.

In August that year, the Soviet Union exploded its first atomic bomb, much to the pride of its citizens and to the surprise of the United States. A nuclear arms race had begun and the following January US president Harry Truman announced to Congress that the country's nuclear scientists would accelerate efforts to develop a more powerful fusion weapon, the H-bomb. Lavrentyev realised that it was time to act. He wrote a short letter to Comrade Stalin explaining that he had worked out how to make a hydrogen bomb and also a way to control fusion reactions to generate electricity for industry. He waited. Nothing happened.

A few months later Lavrentyev wrote another letter, this time addressed to the central committee of the Communist Party of the Soviet Union. Then things started happening very fast. An officer came to interrogate him. He was then put in a guarded room and given two weeks to put his ideas down in writing. These were sent to Moscow via the Communist Party's secret courier service on 29th July, 1950. He was then demobilised and sent off to Moscow. Stopping en route in Yuzhno-Sakhalinsk, the island's capital, Lavrentyev was warmly welcomed by the regional Communist Party committee. Following his arrival in Moscow on 8th August, he sat the entrance exam and was enrolled as a student at Moscow State University. He had gone, in just a few weeks, from lowly radio operator in the far east to a student at the Soviet Union's elite research university in Moscow.

In September he was summoned to see I. D. Serbin, head of the Communist Party's department of heavy engineering industry. Serbin again asked Lavrentyev to write down his ideas about thermonuclear electricity generation, which he did in a classified security-protected room. Lavrentyev settled into the life of a Moscow university student but one evening the following January, as he arrived back at his room in the student hostel, he was told to ring a certain phone number. The person who answered was V. A. Makhnev, the Minister of Measuring Instrument Industry (the codename for the Soviet nuclear industry). Makhnev told him to come immediately to his office in the Kremlin. A man met Lavrentyev at the security gate and walked with him to Makhnev's office. The minister then introduced Lavrentyev to his guide, Andrei Sakharov, father of the Russian H-bomb and, much later, a prominent Soviet dissident.

The previous summer, while working at Arzamas-16, one of the Soviet Union's secret nuclear cities, Sakharov had been asked to look at the paper Lavrentyev had written in Sakhalin. In his report, Sakharov dismissed Lavrentyev's idea for an H-bomb with a single sentence. Lavrentyev had proposed fusing hydrogen and lithium but Sakharov pointed out that this combination was not reactive enough to work in an atomic explosive (something that Lavrentyev would not have known from his readings on Sakhalin). The second suggestion, of using controlled fusion reactions to generate electricity, Sakharov found much more interesting. 'I believe the author has formulated an extremely important and not necessarily hopeless problem,' he wrote.

Lavrentyev's scheme used electric fields to confine a plasma so that it would fuse. He described having two concentric spheres: the outer one would act as an ion source while the inner one, made of a metallic grid, would be put at a large negative voltage with respect to the outer one. This creates an electric field that will accelerate deuterium ions from the outer sphere towards the centre, where they form a hot plasma and fuse. The field also prevents the ions from escaping from the sphere. Sakharov pointed out a couple of reasons why the device, known as an electromagnetic trap, wouldn't work. First, while the electric field would indeed propel ions towards the centre of the sphere, it works in the opposite direction for electrons, so they would be ejected from the device. As a result the plasma in the centre would be dominated by a positive electric charge which would stop the nuclei getting close enough together to fuse. Sakharov also thought that the low density of the plasma would mean that not enough collisions would take place. He didn't exclude the possibility that improvements to the design could overcome these problems, but he concluded by saying, 'It is necessary at this point not to overlook the creative initiative of the author.' Lavrentyev, he thought, was worth cultivating.

Makhanev told the two of them that they would need to meet the chairman of the Special Committee on atomic weapons. Some days later they were again ushered into the Kremlin and to the chairman's office. Sitting behind the desk was Lavrentiy Beria, the most feared man in the Soviet Union. Beria was in charge of Soviet security and the notorious secret police of the NKVD, the forerunner of the KGB. During the war he coordinated anti-partisan operations and ordered the execution of thousands of deserters and 'suspected malingerers.' He hugely expanded the Gulag slave labour camps and ordered the Katyn massacre of 22,000 Polish army and police officers. After the war he was elevated to deputy premier and took personal control of the crash programme to develop nuclear weapons. With Lavrentyev and Sakharov he was polite and formal, asking them about their families, including relatives who were in prison. Nothing nuclear was discussed. Lavrentyev got the impression that Beria was sizing them up, assessing what sort of people they were. As they left, Sakharov said to Lavrentyev that from then on everything would go smoothly and they would work together.

For Lavrentyev things did go smoothly. One evening, much to the alarm of his fellow students, darkly dressed men came and took him and his belongings away in a black limousine. His classmates feared the worst but he was back in lectures the next day having been installed in his own furnished room near the city centre. He was also given a generous scholarship, delivery of any scientific literature he asked for, and the university's professors of physics, chemistry and mathematics tutored him personally. Soon after his interview with Beria he was visited by another official who took him to a government building where, after many security checks, he was introduced to two generals and a civilian with a copious dark beard. This time the conversation was much more technical but when it moved on to his H-bomb design Lavrentyev was unsure whether he should be talking about it. He told the three men that he had recently been to see Beria and the talk veered towards more practical matters. The bearded civilian was Igor Kurchatov, the head of the Soviet nuclear weapons programme. During the war, Kurchatov had vowed not to cut off his beard until his programme to develop a nuclear bomb for the Soviet Union had succeeded. After the first bomb was tested in 1949, Kurchatov decided to keep the beard and he often wore it trimmed into eccentric shapes. To his loyal staff, he was known simply as 'The Beard.'

In May 1951 Lavrentyev was granted clearance to work at the Laboratory of Measuring Instruments (LIPAN), the Soviet Union's secret nuclear research laboratory, alongside his university studies. When he arrived there, he found that a high-powered team was already working on a fusion device. To his dismay, it was not his electromagnetic trap but a magnetic design devised by Sakharov and his colleague and mentor Igor Tamm.

Despite now being at the very heart of the Soviet Union's nuclear research effort, a place he could only have dreamed about a year earlier, Lavrentyev never fitted in there. His connection with the hated Beria made him an object of suspicion, while at the university his privileged status set him apart from the other students.

After the death of Stalin in March 1953, there was a brief struggle for power among members of the Politburo. Most of them feared and distrusted Beria and in June they had him arrested, moving a tank division and a rifle division into the city to prevent security forces loyal to Beria from rescuing him. Six months later Beria and six accomplices were accused of being in the pay of foreign intelligence agencies and of wanting to restore capitalism. They were executed days later. With his sponsor gone, Lavrentyev's privileges evaporated and he was barred from LIPAN. Nevertheless, he finished his physics degree and went on to complete a doctorate, even without access to a laboratory. He got a job at the Physical-Technical Institute in Kharkov, Ukraine, where he continued, without success, to work on plasma confinement with an electromagnetic trap. Sakharov stayed in contact with Lavrentyev and always insisted that it was Lavrentyev's paper written on Sakhalin that provided the spark for Russia's controlled fusion research.

When Sakharov received that paper in the summer of 1950, he had already been thinking about controlled nuclear fusion but couldn't figure out a way to achieve it. Although he didn't think Lavrentyev's electromagnetic trap would work, it did get him wondering about a magnetic trap instead. He discussed the problem with Igor Tamm. Although the two of them were working frantically on the H-bomb, they took some time out to consider the problem of controlled fusion. Following the same thought processes that Lyman Spitzer would work through in a few months' time, they realised that charged particles could be held in place by a magnetic field because they would be forced to move in tight circles around the magnetic field lines. And by making the field lines go around in a circle inside a toroidal tube they could avoid the particles escaping off the end of the field lines – the lines would have no end. Like Spitzer, Sakharov and Tamm realised that the fact that the magnetic field was stronger on the inside of the curve than the outside would push particles towards the outer wall. But while Spitzer got around this by twisting the tube into a figure-of-8, the Russians instead twisted the magnetic field. They added a second magnetic field

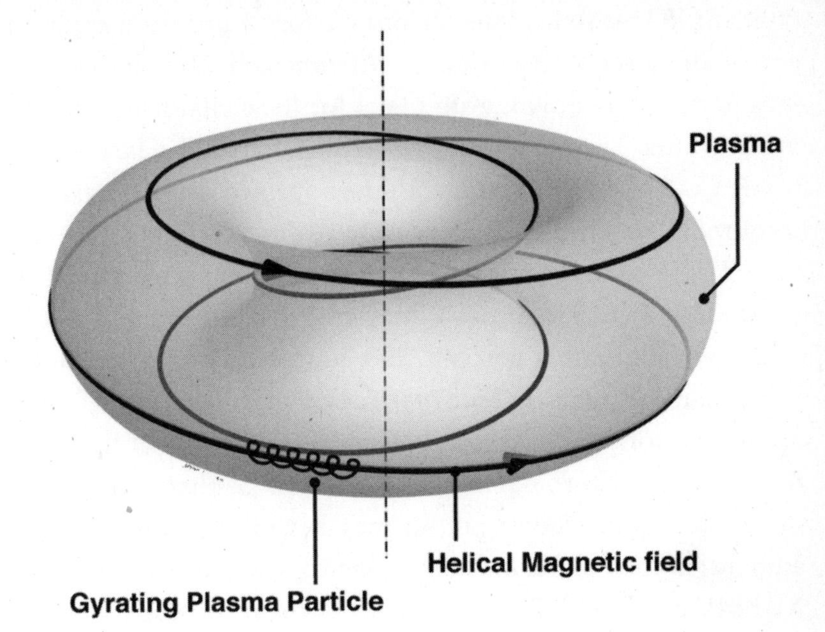


**In a tokamak, plasma particles still gyrate around the magnetic field lines but the lines also follow a helical path around the torus.** (Courtesy of EFDA JET)

that did a vertical loop around the inside of the tube, and the combination of the two fields led to field lines that wound around the torus in a helical pattern. So a particle travelling around the ring would move in a tight spiral around a field line, and that field line would curve down the outer wall, say, at an angle and then across the bottom and up the inner wall. Hence the tendency of particles to drift towards the outer wall would be cancelled out by visits close to the inner wall.

Sakharov and Tamm enlisted the help of some theoreticians from the Lebedev Physics Institute in Moscow to flesh out the idea and then presented it to Kurchatov. Their boss was enthusiastic about the idea and began assembling a group of physicists to work on it at LIPAN. To lead the scientific effort, Kurchatov appointed one of his most able deputies: Lev Artsimovich. On 5th May, 1951 – as Spitzer was drawing up plans for his stellarator and a few months after Thonemann moved from Oxford to Harwell – the USSR Council of Ministers, with the approval of Stalin, passed a resolution launching Russia's fusion research programme.

The researchers at LIPAN first turned their attention to how to produce a current around the torus. Starting off with a high-frequency alternating current, they soon switched to a unidirectional current pulse. In fact they found that the current moving around the torus created such a strong pinch effect that they began to wonder if the toroidal magnetic field that Sakharov had started out with was needed at all. So they began experimenting with straightforward pinch machines akin to Thonemann's Mark I to Mark IV and Tuck's Perhapsatron. But the researchers were having trouble getting higher temperatures from the pinch machines and although they produced some neutrons in July 1952 these turned out to be spuriously produced by instabilities.

As a result, their approach swung back towards using a combination of toroidal and poloidal magnetic fields, just as Sakharov had suggested. In 1955 they built their first tokamak-like device, although they hadn't yet given it that name. However, like its predecessors, the new machine was not producing high temperatures. They didn't realise it until later, but the problem was that the Russians were making their reactor vessels out of ceramics. Atoms knocked out of the vessel's walls were polluting the plasma and radiating energy out of the plasma as UV light, so preventing it from getting hot. The Russian effort was not making progress and they were running out of ideas.

Partly as a way of injecting some new blood into his fusion teams, Kurchatov decided in 1955 that he needed to get the field declassified. He started by organising a conference of scientists from all over the Soviet Union and revealed LIPAN's work on controlled fusion. Delegates, who didn't even know that the programme existed, were stunned by the scale and quantity of the work already done. The following April, Kurchatov surprised western scientists by describing Soviet fusion research in his famous speech at Harwell. Now that the ice was broken, discreet connections between East and West were made at scientific conferences, even though the topic had not been officially declassified. In the autumn of 1956, for example, Artsimovich and a colleague attended an astrophysics conference in Sweden where they made the acquaintance of Princeton's Lyman Spitzer and Harwell's Sebastian Pease.

With their own research stalled, the Russian team was astonished to read in British newspapers in January 1958 about the success of ZETA. From the scant details and photographs in the media reports, theorists at LIPAN scrambled to figure out what sort of a machine ZETA was. From the pictures, they realised that it had to be a compact torus – more like a doughnut than a hula-hoop – but the only way they could see to contain the plasma in such a vessel was with tokamak-like magnetic fields. Once the issue of *Nature* arrived with the articles by the ZETA team and their US colleagues, the Russians realised they were wrong. Although the ZETA results turned out to be false, the theoretical work that the LIPAN team did trying to understand it helped them in the plans for the first large tokamak, T-3.

In Russia just as in the West, nuclear authorities decided to completely declassify Soviet fusion research just before the 1958 Geneva conference, so the Russian researchers arrived there with all their papers recently published in hefty four-volume sets, ready to share with their new-found Western colleagues. Spitzer's stellarators were the surprise of the conference for the Russians. It was an approach, with its steady-state operation and long figure-of-8 or racetrack-shaped vessels, that had not occurred to them. Kurchatov was so taken with the idea that he ordered construction of the T-3 to stop so that a stellarator could be built in Moscow instead. LIPAN theorists compared the two approaches and argued forcefully in favour of the tokamak. Kurchatov relented and they left the stellarator to the Americans.

Russian fusion research was still having problems, but at least they were problems that could now be shared with like-minded colleagues in the West. The challenges they all faced included short confinement times, low plasma temperatures and Bohm diffusion. Following the ZETA debacle, researchers everywhere were going back to basic plasma physics and trying to understand better how plasmas work. The Soviet authorities considered the pursuit of fusion a high priority so there was no shortage of money for experiments at LIPAN. Because their understanding of plasma was poor and their instruments were rudimentary, the team of mostly young researchers there would simply build one device after another with slight variations of design to try out different ideas in the hope that one might show some improvement in performance.

One young researcher, Vladimir Mukhovatov, was trying to stop oxygen, absorbed in the walls of his device, from leaking into the plasma and contaminating it. He decided the best material to coat the inside surface would be gold. He knew that such an experiment would be hugely expensive but nonetheless he went to his team leader, Nutan Yavlinskii, and explained the idea. A week later a 2 kilogram lump of gold was sitting on his desk. Mukhovatov coated the inside of the device with gold, but for some reason it only made its performance worse: he couldn't get a good density and it was plagued with instabilities. When he looked inside he found flakes of gold falling off the walls. He removed the gold but it was now mixed with all sorts of other material, so Mukhovatov sent the mixture off to a special workshop in the Urals to have the gold extracted. Months later he phoned the workshop and was told that they had only found traces of gold in the material he had sent, nothing like the 2 kilograms he was expecting – someone there had seen a golden opportunity and spirited the metal away. Mukhovatov nervously went to Yavlinskii to confess his mistake but his boss dismissed it with a wave of his hand, as if to say there's more where that came from.

Three years after the Geneva conference came the IAEA's next meeting of fusion researchers in Salzburg, Austria. It was here that Western researchers got their first blast of Lev Artsimovich. As leader of the Soviet fusion programme Artsimovich was a force of nature: he knew every device and every theory back-to-front and was constantly analysing data, assessing models and throwing out new ideas. He was the heart and soul of the fusion effort at LIPAN. In seminars he sat in a shabby oak armchair which, legend has it, once belonged to the famous quantum theorist Werner Heisenberg but had been 'liberated' from the Kaiser Wilhelm Institute in Berlin by Russian soldiers at the end of the war. Artsimovich would listen intently and always had the knack of identifying key points and possible weaknesses straight away. After such talks, young researchers and seasoned veterans alike would cluster around the blackboard and debate the issue vigorously with the chief. To a newcomer, used to the customary stuffy formality of Russian science, it was an exhilarating experience.

At the Salzburg conference, Artsimovich lambasted the optimistic results presented by Dick Post of Livermore, who was reporting long confinement times in a magnetic mirror machine, much longer than that reported by M. S. Ioffe of LIPAN using a similar device. 'I want to say that Ioffe's results are in sharp contradiction with the attractive picture of a thermonuclear Eldorado . . . drawn by Dr Post,' Artsimovich taunted. It turned out that, because of a mistake in interpreting measurements, Livermore's results weren't nearly as good as reported and Artsimovich made sure that everyone's attention was drawn to this mistake.

Artsimovich and his team continued to work on their tokamak design. Elsewhere, such pinch-based machines had fallen out of favour. Tuck at Los Alamos had lost interest and really only the British were continuing with pinches, although ZETA continued to be plagued by instability. But the Russian tokamak, with its pinch reinforced by a strong longitudinal magnetic field as a backbone, continued to improve. The researchers at LIPAN, now known as the Kurchatov Institute of Atomic Energy following the death of The Beard in 1960, developed new diagnostic techniques to study the plasma and ways of controlling it.

By the time of the next IAEA fusion conference in 1965, held at Culham, Artsimovich had some impressive results to report: plasma temperature of 1 million °C and confinement time between 2 and 4 milliseconds, ten times better than Bohm's formula predicts. As we heard previously, the Russian results generated a lot of interest at the Culham conference, especially since the tokamak was almost unknown outside the Soviet Union, but Lyman Spitzer and others remained sceptical because of the indirect measurements of temperature and confinement.

Undaunted, the Russians continued to work on their tokamaks and by the time they hosted the IAEA conference at Novosibirsk in Siberia in 1968, their results could not be dismissed so easily. Now Artsimovich was boasting a temperature for the electrons in the T-3 tokamak of 10 million °C and a confinement time of 10 milliseconds, fifty times Bohm's prediction. The meeting was buzzing with the news. There were other machines that were breaking the Bohm barrier, but it seemed that with a tokamak you could improve its performance further by making it bigger and strengthening its magnetic fields. Here, at last, seemed to be a device that would allow fusion researchers to move forward and create plasmas at thermonuclear temperatures.

There were still doubters, however, principally from Princeton. They argued that the byzantine method the Russians used to estimate the electron temperature left it open to doubt. The Russians derived the overall temperature of the plasma (ions and electrons) by measuring its magnetic properties. The ion temperature is taken from the energy of certain ions that get neutralised and then ejected from the plasma. To get the electron temperature the Kurchatov team subtracted the second measurement from the first. The Princeton researchers argued that so-called runaway electrons, of the sort that gave misleading neutron readings in ZETA and the Perhapsatron, could also be confusing the Russian measurements.

The lack of effective measurement techniques was a problem that plagued the early decades of fusion research. Not knowing what the plasma was doing made it much harder to see how to improve its properties. A few years after the Novosibirsk conference, during a debate on fusion research in the British House of Lords, one peer asked: 'How do they measure a temperature of 300 million °C?' The answer offered was: 'I expect that they use a very long thermometer.' If you did try to measure such a temperature with a standard mercury-in-glass thermometer, it would have to be around 600 kilometres long. Even if that were possible, the enormous temperature of the plasma would simply melt the glass.

The correct response to the lord's question would have been, not a very long thermometer but a laser beam. Researchers at Culham had begun five years earlier to try to measure plasma temperature using lasers. Lasers had only been invented a few years previously in 1960 but researchers quickly realised how useful they could be. One of the key aspects of laser light is that its photons all have exactly the same frequency. This is useful because if you shine a laser beam at, say, a plasma of rapidly moving particles, some of the photons will be scattered by collisions with the particles. A photon that collides with a particle moving towards it gets a slight energy boost which increases its frequency. The faster the colliding particle, the greater is the shift in frequency. A photon colliding with a particle moving away from it loses some energy, lowering its frequency. These examples of the Doppler effect can be put to good use: if you fire a laser beam into a plasma and analyse the frequencies of the scattered photons, a small spread of frequencies suggests that the plasma particles were not moving very fast (i.e., had a low temperature) because the shifts in frequency were small; a high temperature plasma, with faster moving particles, would smear out the frequencies across a broader range. So this technique of 'Doppler broadening' of the scattered photons can be used to measure the plasma's temperature.

By 1968, Culham researchers had shown that they could measure plasma temperatures in their pinch machines much more accurately than with the indirect methods used before. This capability would be invaluable to Artsimovich because he would be able to prove the achievements of his tokamak. So it was now that he proposed to Pease that the British should send a team of researchers from Culham to Moscow to settle the question. This was revolutionary. It was the height of the Cold War and some of the technology needed for the experiment could be militarily sensitive. The year 1968 was, however, a unique moment in history. At the start of the year, Czechoslovak leader Alexander Dubček began a process of liberalisation in his Eastern Bloc country, restoring freedom of speech and travel, loosening ties on the media, decentralising the economy and promoting democracy. All across Eastern Europe there was hope that the Soviet Union's stranglehold on their countries might be loosened. Elsewhere there were similar upheavals: France was brought close to revolution by rioting students; in the US the Civil Rights Act was passed and demonstrators marched against the Vietnam War. Change and opportunity were in the air and this must have spurred on the two physicists to push for this unprecedented project.

Pease had to pull every string available to get approval, working his way up to the management of the UK Atomic Energy Authority and on to the Ministry of Technology and the Foreign Office. His ace card was the fact that the mighty Soviet Union was calling on British expertise to solve its technological problem. In the midst of this negotiation, the Soviet Union decided that the Czech experiment had gone far enough. On 20th August, Russian and other Eastern Bloc tanks rolled into Czechoslovakia, Dubček was removed and the Prague Spring was brought to an abrupt end. The familiar Cold War animosities were resumed and this delayed approval of the Culham scientists' mission. Finally in December, an official invitation from the Soviet State Committee for Science to visit the Kurchatov Institute was received, authorised by Soviet premier Leonid Brezhnev. The necessary visas were delivered to the homes of team members at midnight, just hours before their flight, by black-clad motorcycle couriers from the Foreign Office.

For the British fusion scientists – Nicol Peacock, Michael Forrest, Peter Wilcock, and Derek Robinson – who were used to the familiar comforts of the Oxfordshire countryside around Culham, 1960s Soviet Moscow was an alien world. Driving into the city in a chauffeur-driven government limousine with -30°C temperatures outside, they passed the towering pinnacles of Stalinist-era office buildings, as well as the tangled metal fortifications which helped stop Hitler's armies from entering Moscow. This initial visit to Moscow was a fact-finding mission to see the Russian setup and decide what they would need to bring to make the measurements. The tokamak they were to study, the T-3, despite containing probably the highest temperature on Earth, was an unprepossessing sight, a tangle of pipes and wires and unfinished metal surfaces. There were all sorts of problems, including the wildly fluctuating local power supply, vibrations from the giant flywheel generators that powered the tokamak, and stray electric and magnetic fields that would affect the laser.

Back at Culham they had three months to get a suitable laser ready, build the necessary optical equipment and find the right light detectors to pick up the scattered photons. Pease put all the facilities of Culham at their disposal and by mid April 1969 they were ready to go with twenty-six cases full of equipment weighing 5 tonnes. A few items – whose descriptions were deliberately vague in the official inventory – were in fact military-grade light detectors called photomultipliers which were on a list of equipment that it was forbidden to export to communist countries. Another item – a room-sized metal cage to keep stray fields away from the equipment – was so large that the team had to travel in an adapted Boeing 707 belonging to Pakistan Airlines, the only civilian aircraft that regularly flew to Moscow that had big enough doors.

Assembling the equipment in a cramped cellar room underneath the T-3 tokamak – so they could shine the laser up through the underside of the torus – took weeks of intensive work. Artsimovich often came to check on progress, eager to see his tokamak vindicated. Living in Soviet Moscow was not easy for the team members. Local food shops were as bare as those in Britain during the war, although the Britons could supplement from the 'Berioska' shops, only accessible to foreigners with hard currency. Robinson's wife Marion, who had taken leave from her own job as a chemist at Harwell to join the party, did much of the work of finding out how to survive in Moscow. They lived in the same apartment building as many of the young researchers from the Kurchatov institute and forged friendships that lasted decades. It was not all work for the Culham team: thanks to the connections of the Kurchatov representatives of the Communist Party, they were taken to see ballet at the Bolshoi Theatre, opera at the Kremlin Theatre, the czarist crown jewels in the Kremlin Armoury, and the Moscow State Circus.

Once the experimental setup was all in place, they tried for the first time to shine laser light into the plasma and measure the scattered photons. The researchers immediately found that the plasma itself was giving off so much light – much more than they had expected – that it swamped the rather faint scattered photons. For weeks they tried to tease out the signal of the scattered beam without success, with Artsimovich looking anxiously on.

In June they decided they had to implement their backup plan. They had prepared and packed a second laser that produced much shorter pulses. If they illuminated the plasma very briefly with one of these short pulses and then only opened up the detector just long enough to catch the scattered laser photons, they would not catch so much light from the plasma at the same time. They had this backup laser quickly shipped to Moscow and began making the necessary changes. Work continued into July in the sweaty heat of the Moscow summer. The new higher power laser damaged other optical components, so replacements had to be shipped out quickly in deliveries to the British Embassy to avoid delays in customs.

On 21st July, as the rest of the world watched with baited breath for Neil Armstrong to clamber down a ladder onto the surface of the Moon, the Culham team made their final adjustments. The next day, while the Apollo 11 crew were still on their way home, Robinson made a call to Pease at Culham. He said that they had seen a clear signal of scattered laser photons with the new setup and the Doppler broadening suggested that the temperature was high. Another two weeks of experiments and they were sure that the T-3 was achieving temperatures of more than 10 million °C, just as the Russians had said a year earlier in Novosibirsk. The researchers called Culham again and, now that he was sure, Pease telephoned Harold Furth, research director of the Princeton fusion lab in the United States – a series of phone calls that completely changed the course of fusion research.

In the United States, the vindication of the tokamak prompted a variety of reactions. In Princeton it caused despondency. In the offices of the fusion section at the Atomic Energy Commission (AEC) in Washington, DC, staffed danced on the tables. The problem was that the US had invested heavily in fusion. It was now funding research at four different laboratories – Princeton, Los Alamos, Livermore and Oak Ridge. Researchers were investigating a variety of devices, including stellarators, pinches, mirror machines, more exotic geometries called multipoles, and others. Some people had devoted their working lives to these machines. But none of them was performing anywhere near as well as the Russian tokamaks.

At Princeton, the Model C stellarator was not living up to its promise. There was duplication of effort between Princeton and Livermore, and Congress was looking for budget cuts to help finance the war in Vietnam. Scientists, following the example of Lyman Spitzer, were beginning to quit fusion for other fields because the prospect of rapid progress towards power-producing reactors seemed to have evaporated and there were few new ideas pointing to a way forward. The AEC's Amasa Bishop, who now headed the fusion section, needed something to kick-start the US fusion programme, something to enthuse both Congress and his own researchers.

Bishop had visited Russia in 1967 and had been impressed by the machines at the Kurchatov Institute. When Artsimovich presented his startling tokamak results at the IAEA conference in Novosibirsk, Bishop began to think seriously about whether the US needed to start building tokamaks. Researchers from Princeton, of course, dismissed the idea. They thought the Russians were mistaken and that in reality the performance of the tokamaks was not that much better than the Model C. Bishop took their views seriously. They had devoted nearly two decades to developing the stellarator and were America's undisputed experts on toroidal fusion devices.

Others were not so negative, however. The fusion team at Oak Ridge National Laboratory in Tennessee had been struggling for years with an unusual magnetic mirror device called the Direct Current Experiment (DCX) which heated a plasma by firing a beam of deuterium molecules ($D_2$) into it. But after years of trying they could only get it to work at very low particle density. Put too many particles into the plasma and instabilities broke it apart. Oak Ridge's fusion chief Herman Postma feared that, with Congress looking for savings, his lab might be cut out of the fusion programme altogether. What they needed was a new device to rally around and the tokamak seemed to offer the perfect opportunity: no one was building tokamaks in the US and Oak Ridge could become the national tokamak lab. Early in 1969 Postma's team began designing its own Oak Ridge Tokamak, or Ormak, which would aim to both replicate the Russian results (this was before the Culham team had announced their temperature measurements) and to go beyond them to demonstrate something new. The Russian team had shown that plasma performance improved if the ratio of the radius of the torus over the radius of the plasma tube, known as the aspect ratio, was low – in other words, if the tokamak was more like a doughnut than a hula-hoop. So the Oak Ridge team designed Ormak with two interchangeable plasma vessels, one with an aspect ratio equivalent to the Russian T-3 (roughly 7) and another with the much smaller ratio of 2.

In the spring of 1969, Artsimovich came to Boston. He had been invited by two professors at the Massachusetts Institute of Technology (MIT) whom he knew and the visit was meant to be partly a holiday: he would give a few lectures and work on a book he was writing. But with US interest in tokamaks high, researchers just wouldn't leave him alone. Some of the Oak Ridge team designing Ormak came to Boston for a private audience. Another team came from the University of Texas, where they were designing a tokamak with a novel feature. They planned to use a strong electric field to deliberately cause turbulence in the plasma in the hope that eddies would boost its temperature.

Bruno Coppi, an Italian physicist who had recently arrived at MIT, also sought out Artsimovich. Coppi had spent some years working in Princeton and came to MIT with yet another plan for a tokamak. The pinch effect in tokamaks relies on a current flowing around the toroidal chamber. That current has a beneficial side-effect: it also heats the plasma because there is resistance to the flow of current, so when the current is pushed through hard the resistance raises its temperature. It is similar to the way friction between your palms warms your hands when you rub them together. Coppi sought to design a tokamak with a low aspect ratio and a very strong toroidal magnetic field, both of which were thought to maximise this resistance effect, known as ohmic heating (the ohm is a unit of resistance). MIT had a world-class magnet laboratory and Coppi enlisted the help of its engineers to design his machine.

What all these tokamak plans lacked was money to build them, but that was soon to change. Bishop knew that the time was right to move into tokamaks. Even Congress had got interested in the Russian results and was enquiring of the AEC what it needed to catch up with the Soviets. So Bishop arranged a meeting of his standing committee of fusion advisers in Albuquerque, New Mexico, in June 1969 and called for any researchers with tokamak proposals to come and present them. The Oak Ridge team came and proposed Ormak; Texas researchers put forward the Texas Turbulent Tokamak and Coppi and his MIT colleagues suggested their compact high-field machine, known as Alcator, derived from the Latin phrase for high field torus. What Bishop wanted but didn't get was a proposal from the Princeton Plasma Physics Laboratory, or PPPL. They were the torus experts and Bishop figured that the fastest way to duplicate the Russian results in a similar machine would be to cannibalise Princeton's Model C stellarator. Although the Model C was racetrack shaped, if you removed the straight sections and added an electromagnet you would get a tokamak of roughly the same size as Russia's T-3.

But Princeton wasn't playing ball. PPPL researchers insisted at the meeting that the Russian results were mistaken and this led to heated argument. They were, however, fighting a losing battle. After days of constant pressure from the standing committee, PPPL director Mel Gottlieb finally caved in and agreed that it was vital to test the abilities of tokamaks and that Model C should be sacrificed for the purpose. The confirmation of the T-3 temperature by the Culham team a few weeks later sealed the Model C's fate. But the AEC's standing committee was limited by its budget and could only give two projects the green light: the converted Model C and Ormak.

Once the decision was made, the Princeton researchers didn't waste any time. By September they had come up with a design. The plan was to increase the radius of the plasma tube – and so reduce the aspect ratio – and also to cut down the straight sections of Model C's racetrack shape to 20 centimetres. But they learned from the Russians that having a uniform symmetrical magnetic field was important, so the design was changed again and the 20-centimetre straight sections removed altogether. This provided the proposed machine with a name: the Symmetric Tokamak, or ST. Once all the new components were ready, the Model C stellarator was switched off for the last time on 20th December, 1969 and in little more than 4 months the team transformed it into a tokamak.

Unlike the Russian machines, the ST was bristling with measuring devices. It was an unwritten law at the Princeton lab: everything has to be measured. Once they applied those instruments to the plasma in the completed ST they found that everything the Russians had said was true. The temperatures and confinement times were higher than anything yet achieved in the United States. US fusion scientists fell head over heels in love with the tokamak. In 1971, just a year after ST started operating, Oak Ridge's Ormak was ready to go, soon followed by the Texas Turbulent Tokamak. The following year saw the completion of MIT's first machine – Alcator A – while General Atomics in San Diego built an unusual tokamak with a kidney shaped cross-section called Doublet II, and Princeton built a second, the Adiabatic Toroidal Compressor or ATC, to test methods for heating the plasma. More joined them in the following years.

The United States was not the only country to jump on the tokamak bandwagon. Researchers at Culham were in the middle of building a new stellarator called CLEO and, as happened to Princeton's Model C, it was rebuilt as a tokamak. France entered the fusion race by building, straight off, the top performing tokamak of the early 1970s. The Tokamak de Fontenay aux Roses (TFR) had the strongest toroidal magnetic field of the time and could generate a plasma current of 400,000 amps. Germany built a smaller machine, the Pulsator, at its fusion lab in Garching near Munich and the Italians built the small TTF in Frascati. Japan too joined in with its first tokamak, the JFT-2. And the Russians, not keen to relinquish their lead, built a string of new machines to test different ideas.

For researchers outside Russia, the arrival of the tokamak transformed the field. Throughout the 1960s Bohm diffusion had frustrated all their efforts, leeching energy and ions out of the plasma and making high temperatures impossible to reach. Their work had become an effort to understand the behaviour of plasma, rather than a race to produce a power-producing fusion reactor. But the tokamak seemed to show them a way forward to higher temperatures and longer confinement. True, they didn't really have a thorough understanding of how it worked or why it was better than other devices but that would come, they believed, as they got to know the machine better. The important thing was that the race towards a fusion power reactor was back on.

Not that everything was plain sailing. Once researchers started to push the tokamak to the limits, its own menagerie of plasma instabilities began to make an appearance. The potentially most serious was dubbed simply a 'disruption.' In a disruption, the plasma current which does most of the work confining the plasma suddenly collapses to zero in a tiny fraction of a second, leading to a loss of plasma and temperature. More serious than that is the fact that the sudden loss of that huge current (in the millions of amps for a large tokamak) induces strong eddy currents in the vacuum vessel which put it under huge mechanical strain – equivalent to hundreds of tonnes for a big machine. Researchers were obviously keen to avoid such events because of the damage they could potentially inflict on their precious devices. Experiments showed that high plasma pressure, high current and impurities in the plasma, increased the chances of a disruption. They were soon able to draw up a diagram showing the boundaries of safe operation but such boundaries were a problem because high pressure and current are desirable if you want to get to the conditions needed for fusion. Finding ways to push back those boundaries became a high priority.

Another puzzling instability was discovered in Princeton's thoroughly instrumented Symmetric Tokamak. Using an x-ray detector, the PPPL researchers detected an x-signal coming from the plasma that went up and down in a regular repeating pattern; on a plot it looked like sawteeth. The sawteeth were caused by rapid heating and then cooling of electrons in the core and they were also seen in Russia's T-4 (an upgrade of T-3) and Princeton's ATC. Sawtooth instability soon came to be considered a sign that a tokamak had reached respectable operating conditions and its core was good and hot. The Russian researcher Boris Kadomtsev developed a theory that seemed to explain the oscillations and, as they were relatively benign, they were considered in the 1970s as a success story of plasma theory. But as machines got bigger, so did the sawteeth to the point at which they were causing turbulence that spoiled confinement.

But perhaps the most important upshot of the boom in tokamak building during the 1970s was that it allowed researchers to develop formulas that linked plasma properties to tokamak size, known as scaling laws. Even if you don't have a detailed knowledge of how a plasma works, scaling laws allow you to predict what sort of tokamak will give you the best results. It worked because there were so many tokamaks of different shapes and sizes that physicists could operate with different sets of conditions. By plotting the results from many machines on a graph of, say, confinement time against the major radius of the tokamak, you can plot a line that extrapolates from those results to predict what confinement you would get with an even larger tokamak. Researchers did similar plots of confinement against plasma radius, toroidal magnetic field, plasma current and electron density. Some of the results were surprising: although plasma theory predicted that confinement time would go down as the density of electrons increased, real results from tokamaks showed confinement clearly growing the more electrons there were. Overall, there was one clear message: if you wanted to get closer to the temperature and confinement time needed for a plasma to burn with fusion reactions, then the larger the volume of plasma the better. It was time for tokamaks to get big.

CHAPTER 5

# Tokamaks Take Over

IN SEPTEMBER 1958 PAUL-HENRI REBUT, A YOUNG GRADUATE from France's prestigious École Polytechnique near Paris, arrived for his first day of work as a researcher at the Commissariat d'Énergie Atomique (CEA). He had been hired to join the CEA's new fusion department but as he walked in he found the labs and offices largely deserted. His new colleagues were all in Geneva for the second Conference on the Peaceful Uses of Atomic Energy, the meeting that revealed to the world the secret fusion programmes of Britain, the United States and the Soviet Union. Before that meeting there had been some plasma physics research going on in Europe but, beyond Britain's Harwell laboratory, little that was aimed at fusion energy. The Geneva meeting changed all that. Fusion programmes started up in several countries and, in the same month that Rebut started his new job, European collaboration in fusion began.

A few years earlier, European governments had been debating how to build on the success of the European Coal and Steel Community, the precursor to the European Economic Community (EEC) which later became the European Union. At the time the community consisted of only six members: France, Germany, Italy, the Netherlands, Belgium and Luxembourg. Some of these wanted the community to cover sources of energy other than coal, including the new atomic energy which, because of its high development costs, was a prime candidate for international collaboration. Others wanted to create a single market for goods across the member states. Trying to accommodate these divergent goals in one body was thought too difficult so a compromise was thrashed out. On 1st January, 1958 the coal and steel community was joined by two others, the EEC and Euratom, a body to coordinate the pursuit of atomic energy.

Fusion didn't fit naturally into Euratom's remit because the prospects of fusion power were some way off, so Euratom managers asked CERN, the recently-formed European particle physics laboratory in Geneva, if it would take responsibility for fusion. CERN set up a study group to investigate what fusion research was going on in Europe and what role CERN could play. But the CERN council eventually decided that as the ultimate aim of fusion is to generate energy on a commercial basis, such research was outside the limits of its statutes, which restricted it to pursuing basic science. So the ball was back in Euratom's court.

On 1st September in the halls of the Palais des Nations at the start of the Geneva conference the vice president of Euratom, Italian physicist and politician Enrico Medi, sought out his compatriot and fellow physicist Donato Palumbo and asked him to head up Euratom's fusion programme. Born in Sicily, Palumbo was a talented researcher who had graduated from the respected Scuola Normale Superiore in Pisa and then returned to Sicily to teach at the University of Palermo where he was later made a professor. He specialised in theoretical plasma physics so he was scientifically well qualified for the fusion programme but he was a quiet and unassuming man and so not an obvious choice for the cut and thrust of European politics. Palumbo at first refused, saying he wasn't ready for such a job, but Medi persisted and Palumbo finally agreed.

The fission department of Euratom had a head start on fusion and by far the larger budget. Palumbo was given just $11 million for his first five-year research programme. The fission effort had set about creating a series of Joint Research Centres in various countries where Euratom work would be carried out. With his limited resources Palumbo knew that he couldn't create anything that would rival the national fusion labs, which were then growing rapidly. He also disliked bureaucracy and hierarchies, and so decided to take a different tack. He set out to persuade each national fusion lab to sign a so-called association contract to carry out fusion research agreed collectively within Euratom. It wasn't hard to persuade them because Euratom was offering to pay 25% of the labs' running costs. The first to sign up was France's CEA in 1959 and over the next few years all six Euratom members joined the fusion effort.

Palumbo may not have been a bureaucrat but he was a natural diplomat. Committee meetings at each lab always dealt with science issues first, with a bit of business discussed at the end. The central coordinating committee was delicately named the Groupe de Liaison, because France's proud CEA and Germany's Max Planck Society would not have liked anything that sounded too controlling. This low-key approach made him popular in the growing fusion community and reassured the national labs that Euratom was not planning to take over fusion research.

During the first decade of his tenure at Euratom, Palumbo's fusion programme was mostly focused on understanding the plasma physics of confinement and heating. The labs built a variety of mirror machines, pinches and other toroidal devices. They experimented with heating using neutral particle beams and radio waves. The young Rebut found that the field was so new that there was no one there at the CEA who could teach him about plasma theory so he found what literature he could and taught himself, eventually making important contributions to the understanding of plasma stability. But trained as an engineer as well as a physicist, he soon got involved in designing, building and operating small fusion devices. He built a so-called hard-core pinch, a linear device that relied on a copper conductor along its central axis to carry the current to create the pinch rather than a plasma current. He moved onto toroidal pinches, again with a copper conductor along the axis, and became convinced that only toroidal devices would work because mirrors lost too many particles at their ends. This made him somewhat marginal in the French fusion effort since the CEA's largest device at the time was a mirror machine at its laboratory in Fontenay aux Roses, a Paris suburb.

As a whole, European researchers were suffering the same frustrations as their colleagues in the United States: their machines were plagued by Bohm diffusion, confinement was poor, and funding was slowly dwindling. It was a crisis in Euratom's fission section, however, that put the future of the programme in doubt. From the outset the organisation's aim had been to collaboratively develop a prototype fission reactor that member states could then develop commercially. The favoured design was a heavy water reactor with organic liquid coolant, known as ORGEL. But the commercial appeal of separately developing their own reactors was too much for the member states and in 1968 the project collapsed, prompting savage budget cuts in all parts of Euratom. Palumbo was left with just enough money to pay the staff that Euratom employed in each of the associated labs. For a couple of years the programme stumbled on with a small amount of money from the Dutch government.

Just as the programme was at its lowest ebb, Russia announced its astonishing results with tokamaks at the 1968 IAEA meeting in Novosibirsk. Palumbo was in the middle of drafting a proposal for the next five-year programme but realised that the tokamak results changed everything. He started from scratch and arrived at the June 1969 meeting of the Groupe de Liaison with a new proposal. He suggested that the programme should concentrate its efforts on tokamaks and some other toroidal devices, that an extra 20% funding should be provided by Euratom for the building of any new device, and that a group should be set up to investigate building a large device collaboratively by all the associated labs. There was a heated debate over putting so much emphasis on tokamaks, but they reached an agreement and sent Palumbo's proposal to the Euratom council for approval. Euratom, still smarting from the collapse of ORGEL, enthusiastically endorsed the fusion plan and even gave Palumbo slightly more funding than he had asked for. The laboratories set about drawing up plans for new machines.

The CEA, with its emphasis on mirror machines, was thrown into confusion when fusion fashion switched to tokamaks. Rebut, who had focused on toroidal devices, became the man of the moment. At the time of the Novosibirsk meeting he had been in the middle of designing a large pinch device, but he immediately abandoned it and began working on a tokamak design. When the Groupe de Liaison met again in October 1971, five new machines were approved with the new additional funding. At the same time, a small group was set up to investigate building a large multinational tokamak. This would-be machine was given the name the Joint European Torus (JET) – not 'tokamak' because German delegates thought it sounded too Russian.

Of the machines that were given the go-ahead at that meeting, the most ambitious was Rebut's Tokamak de Fontenay aux Roses (TFR). The design was roughly the same size as Russia's T-3 and Princeton's Symmetric Tokamak but the plasma tube, with a radius of 20cm, was larger so that it contained more plasma and had a lower aspect ratio. What made the TFR stand out was the huge amount of electrical power that was put into containment. Rebut designed a large flywheel that was accelerated up to a high speed over a long period and then acted as an energy store for each plasma shot. Connecting the flywheel to a generator created a huge electrical pulse which, via the tokamak's electromagnet, drove the plasma current that pinched the plasma in place. The TFR was able to generate plasma currents of 400,000 amps for up to half a second, world records at the time.

While the TFR was still being built, the JET study group had to decide what the next generation of tokamaks would be like. They didn't have many practical details to work with because few tokamaks had yet been built outside Russia. The Symmetric Tokamak had started operating in 1970 and Ormak in 1971; in Europe there was only Britain's CLEO, converted from a part-finished stellarator. And they couldn't rely on theory either; there simply was no conceptual understanding of how tokamaks worked. What they did know was that tokamaks got great results and that if you made them bigger their operating conditions would probably get better. The JET study group knew that they wanted a machine that got close to reactor conditions and produced a significant amount of fusion power – and that meant a plasma that would heat itself.

Up until this point, fusion devices were not built to be genuine reactors because they generated so few fusion reactions. Their main aim was to practice confining and heating plasma. Now that devices were getting closer to producing lots of fusion reactions, reactor designers had to take into account the copious energy and particles the reactions would produce. Fusing a deuterium nucleus and a tritium nucleus produces two things. First is a high-energy neutron which has 80% of the energy released by the reaction so it will be moving very fast. Because neutrons are electrically neutral they are not affected by the tokamak's magnetic field and so zap straight out and bury themselves in the tokamak wall or something else nearby, converting their kinetic energy into heat. In a power reactor the idea is to get those neutrons to heat water, raise steam, and use that steam to drive a turbine, turn a generator and create electricity.

The other thing produced in the reaction, carrying the remaining 20% of the energy, is a helium nucleus, otherwise known as an alpha particle. The alpha particle is charged so it will follow a helical path around magnetic field lines in the tokamak just as the ions and electrons of the plasma do. Capturing the alpha particles in the field serves a useful purpose: as they twirl around in the plasma they knock into other particles, transfer energy and generally heat things up. The designers of fusion reactors very much wanted to exploit this effect. If you can get alpha particles to heat up your plasma, that will help to keep the fusion reaction going and may even make other heating methods, such as neutral particle beams and radio-frequency waves, unnecessary.

The difficulty is that the newly created alpha particles are heavier than hydrogen and moving fast, so their spirals will be much wider. If the plasma vessel is too small, the spiralling alphas won't get very far before hitting the wall. A stronger magnetic field will make the alpha particle spirals smaller, and the field is created in part by the plasma current. So reactor designers could derive a formula for successful alpha particle heating: for a given size of plasma vessel, there is a minimum plasma current that will create a strong enough field to keep the alpha particles within that vessel.

The JET study group reported back to the Groupe de Liaison in May 1973 and recommended that Euratom should build a tokamak around 6m across which, to contain alpha particles, would need to carry a current of at least 3 million amps (MA). This was a huge leap from the tokamaks of the day. France's TFR was just 2m across and carried 400,000 amps and it was only just being finished. Physicists had no real idea how plasma would behave with 3MA of current flowing through it. But the excitement about tokamaks was so great in the early 1970s that Palumbo acted on the study group's suggestions straight away and started to assemble a team to work out a detailed design for such a reactor. The obvious man to lead that effort, the man who had just finished building the world's most powerful tokamak, was Paul-Henri Rebut.

In the United Kingdom, Sebastian Pease's Culham laboratory was struggling. Researchers there had kept on working with ZETA during the 1960s and had done some good science. But attempts to build large follow-on machines – ZETA 2 and another one called ICSE – were thwarted by the government. Researchers had to content themselves with smaller devices. As the decade wore on, however, funding to the lab was repeatedly cut. Pease realised that if his lab was ever going to get involved in operating a large device again, it would have to be through collaboration with Britain's European neighbours. But there was a problem: at that time Britain was not a member of the EEC or Euratom. Nevertheless, Pease talked to Palumbo and he allowed Pease and his colleagues to attend meetings of the Groupe de Liaison and join in discussions about JET.

By 1973, when Palumbo was putting together a team to design JET, Britain had joined the EEC and Euratom, and hence Pease was able to offer Culham as a base for the JET design team. So it was that in September 1973, only a few months after the study group proposed that JET should be built, Rebut set sail for England to begin the work. And he did literally set sail. Reluctant to leave behind a yacht that he had designed and built himself but which was not quite finished, Rebut sailed across the English Channel to take up his new job.

Initially housed in wooden huts left over from Culham's time as a World War II airbase, the JET team had a daunting task ahead of them, and just two years to do it in. With the understanding of tokamaks still so sketchy there were two routes they could take: the safe, conservative route of simply scaling up in size from existing, successful machines such as TFR; or the more risky path of trying out some untested ideas. With the flamboyant Rebut in charge it was always going to be the latter. But the designers were constrained by a number of factors: disruptions and other instabilities, for example, limited the plasma's density and the current that could flow through it. There were practical considerations, such as the strength of magnets. And there was cost.

Bearing all of these issues in mind, the team came up with a design for a tokamak that was 8.5m across and the interior of the plasma vessel was twice the height of a person. The volume of the plasma vessel was 100 cubic metres ($m^3$), a vast space compared to TFR's $1m^3$. The current flowing through JET's plasma (3.8MA) would be ten times that in TFR.

Earlier tokamaks – designed principally to study plasma – did not use the most reactive form of fusion fuel, deuterium-tritium. JET would be different but this added a hornet's nest of problems to the project. Tritium is a radioactive gas which, because it is chemically identical to hydrogen, is easily assimilated into the body. So extreme care has to be taken to avoid leaks and every gram of it has to be accounted for. Carrying out deuterium-tritium, or D-T, reactions also produces a lot of high-energy neutrons. When these collide with atoms in the reactor structure it can knock other protons or neutrons out of their nuclei, potentially turning them radioactive. So over time, the interior of the tokamak acquires a level of radioactivity from the neutron bombardment. The levels are nothing like in a fission reactor, but they are enough to make it dangerous for engineers to go inside to carry out repairs or make alterations. JET had to be designed so that any work inside the vessel could be done from outside with a remote manipulator arm, even though such machines were only in their infancy at that time.

Everything about the design was big and ambitious, but one innovation stood out. Rebut decided to make the shape of the plasma vessel D-shaped in cross-section rather than circular. Part of the motivation for this was cost. The magnetic fields required for containment exert forces on the toroidal field coils, the vertical ones that wrap around the plasma vessel. These magnetic forces push the coils in towards the central column of the tokamak and are stronger closer to the centre. JET would require major structural reinforcement to support the toroidal field coils against these forces which would be very expensive. So, Rebut mused, why not let the coils be moulded by the magnetic forces? If left to find their own equilibrium, the coils would become squashed against the central column into a D-shape and stresses on them would be greatly reduced. So Rebut designed a tokamak with D-shaped toroidal field coils and a D-shaped plasma vessel inside them. In cross-section the vessel was 60% taller than it was wide. Overall, the tokamak now looked less like a doughnut and more like a cored apple.

But more importantly Rebut thought a D-shape would get him better performance. In 1972 the Russian fusion chief Lev Artsimovich and his colleague Vitalii Shafranov calculated that plasma current flows best when close to the inside wall of a plasma vessel – close to the central column. So if the plasma vessel was not circular in cross-section but was squashed against the central column, more plasma current would gain from the most favourable conditions. The Russians were in the process of testing the idea but at the time there was little proof that it would work. Rebut was convinced that the key to confinement was a high plasma current and a D-shape, he believed, could allow him to go far beyond the 3MA in JET's specification. He guessed that the Groupe de Liaison would not be persuaded by the Russians' theoretical prediction alone, so adding the issues of cost and engineering would help his case. He was right to be worried. When the JET team presented its design in September 1975, there was vigorous debate about the size of the proposed reactor, the D-shaped plasma vessel, the cost and more besides. Palumbo admitted that he would have preferred the tried-and-tested circular vessel but he trusted Rebut and his team and argued for their design.

Such was the persuasiveness of Rebut that the various Euratom committees eventually agreed to go ahead with JET pretty much as described in the design report. The Council of Ministers, the key panel of government representatives that oversaw the EEC and Euratom, also approved the plan. Constructing JET was predicted to cost 135 million European currency units (a forerunner of the euro) of which 80% was to come from Euratom coffers and the rest from member governments. All that remained was for the council to decide where it was to be built. Although there were technical requirements for the site, this was predominantly a political decision. The fifty-six-strong JET design team waited at Culham to hear the decision. Most of them would move straight to new jobs helping to build JET, so there was little point returning to their home countries.

Hosting such a high-profile international project was viewed as a prize by European nations and soon there were six sites vying for the honour: Culham; a lab of France's CEA in Cadarache; Germany offered its fusion lab at Garching and another site; Belgium offered one; as did Italy with the Euratom-backed fission lab at Ispra. In December 1975 the Council of Ministers debated the site issue for six hours and came away without a decision. So the politicians asked the European Commission, the EEC's executive body, to make a recommendation. The commission opted for Ispra, but when the council met again in March Britain, France and Germany vetoed this suggestion.

The council met again several times in 1976 but still without any resolution of the issue. Rebut and the team at Culham were getting desperate. Some were accepting jobs elsewhere while others were simply returning to their old pre-JET jobs at home. The

European Parliament demanded that the matter should be settled. Politicians began to talk of the project being on its deathbed. The council discussed whether they should abandon their normal rule of making unanimous decisions and decide it by a simple majority, but they failed to decide on that too. All they did manage to do by this time was whittle the list down to two candidate sites: Culham and Garching. The frustrated scientists at Culham sent petitions to the council; their families sent petitions to the council. But by the summer of 1977 the game was up and Rebut and his Culham host Pease decided it was time to wind up the JET team.

On 13th October, while the team was still disbanding, fate intervened. A Lufthansa airliner en route from Mallorca to Frankfurt was hijacked by terrorists from the Popular Front for the Liberation of Palestine. They demanded $15 million and the release of eleven members of an allied terrorist group, the Red Army Faction, who were in prison in Germany. Over the following days, the hijackers forced the plane to move from airport to airport across the Mediterranean and Middle East before finally stopping on 17th October in Mogadishu, Somalia, where they dumped the body of the pilot – whom they had shot – out of the plane. They set a deadline that night for their demands to be met. German negotiators assured them that the Red Army Faction prisoners were being flown over from Germany but at 2 a.m. local time a team of German special forces, the GSG 9, stormed the plane. In the fight that followed, three of the four terrorists were killed and one was wounded. All the passengers were rescued with only a few minor injuries.

So what was the connection with JET? Germany created the GSG 9 in the wake of the bungled police rescue of Israeli athletes after their kidnap by Palestinian terrorists at the Munich Olympics in 1972. For training, the GSG 9 went to the world's two best known anti-terrorist groups: Israel's Sayeret Matkal and Britain's Special Air Service (SAS). Mogadishu was the GSG 9's first oper-

**A home at last: the JET design team, with Paul-Henri Rebut at front centre, celebrates the decision to build JET at Culham, a week after the end of the Mogadishu hijack.**
(Courtesy of EFDA JET)

ation and two SAS members travelled with the group as advisers, supplying them with stun grenades to disorient the hijackers.

The whole hijack drama had been followed with mounting horror, especially in Germany. When the news of the rescue broke on 18th October the country was awash with relief and euphoria, especially when a plane arrived back in Germany carrying the rescued passengers along with the GSG 9, who were welcomed as heroes. Into this heady atmosphere stepped the British Prime Minister, James Callaghan, who arrived in Bonn on the same day for a scheduled summit meeting and was greeted by German chancellor Helmut Schmidt with the words: 'Thank you so much for all you have done.' There were a number of EEC matters that divided Britain and Germany at the time and in the happy atmosphere of that summit meeting many of them were put to rest, including the question of JET's location. A meeting of the Council of Ministers was hastily called a week later. The result was phoned through to Culham around noon and the champagne was, finally, uncorked. At last JET was ready to be built and Rebut and what remained of his team didn't have to move anywhere.

Long before climate change became widely recognised as a threat to our future, an environmental movement grew up in the United States which drew attention to the pollution of the atmosphere by the burning of fossil fuels. This public pressure led to the 1970 Clean Air Act and the creation of the Environmental Protection Agency. At the same time, America's electricity utilities were having trouble keeping up with the soaring demand for power, leading to often serious blackouts and brownouts. The Nixon administration responded by making the search for alternative energy sources, with less environmental impact, a national priority. The Atomic Energy Commission was pinning its hopes on the fission breeder reactor which it had been developing for some time. The AEC argued that the breeder produced less waste heat than the light-water fission reactors that power utilities were building at the time, and burned fuel more efficiently. But the increasingly vocal environmentalists didn't buy it. As far as they could see, breeder reactors had the same safety concerns as light-water machines plus their plutonium fuel was toxic, highly radioactive and a proliferation risk.

With the government looking for alternative energy sources and the public suspicious of nuclear power, fusion scientists suddenly found themselves in demand. Fusion was seen as a 'clean' version of nuclear energy and the idea of generating electricity from sea water seemed almost magical. Researchers from Princeton and elsewhere were now being interviewed by newspapers, courted by members of Congress, and the new tokamak results meant that they really had something to talk about. As if on cue, Robert Hirsch, someone very well suited to make the most of this new celebrity, was put in charge of fusion at the AEC. Hirsch was in his late 30s; he wasn't a plasma physicist, but he was a passionate advocate of fusion and he knew how to play the Washington game: he was at home in the world of Senate committees, industry lobbyists and White House staffers. In 1968 he had been working with television inventor Philo T. Farnsworth on a fusion device using electrostatic confinement and applied to the AEC for funding. Instead, Amasa Bishop hired him. He worked under Bishop and his successor Roy Gould but was always frustrated by the relaxed, collegial approach of the fusion programme. In Hirsch's mind, fusion should be the subject of a crash development programme like the one that sent Apollo to the moon.

In 1971, Hirsch got his chance. Leadership of the AEC changed from nuclear physicist Glenn Seaborg to economist James Schlesinger, then assistant director at the Office of Management and Budget. Schlesinger wanted to counter criticism at the time that the AEC was simply a cheerleader for the nuclear industry; he wanted to diversify into other types of energy. One of his first changes was to promote the fusion section – which at that time was part of the research division – into a division in its own right. Gould, an academic from the California Institute of Technology, stuck to it for around six months and then stood down. Hirsch, like Schlesinger, was interested in planning and effective management. He was a perfect fit and took over as head of the fusion division in August 1972.

At that time, the fusion division was a tiny operation: just five technical staff and five secretaries. Its role and operations had not changed much since the 1950s. The direction of research and its timetable was pretty much decided by the heads of the labs. The fusion chief at the AEC acted as referee between the competing labs and was their champion in government. Hirsch had very different ideas. First he wanted more expertise in the divisional headquarters so that decisions about strategy could be made there. Within a year he had more than tripled the technical staff and by mid 1975 the division boasted fifty fusion experts and twenty-five support staff. He created three assistant director posts in charge of confinement systems, research, and development and technology. Now the lab heads had to report to the various assistant directors, not to Hirsch himself.

Hirsch also wanted the programme to be leaner and more focused, and that meant closing down some fusion devices which were not helping to advance towards an energy-producing reactor. Before the end of 1972 he closed down two projects at the Livermore lab: an exotic mirror machine called Astron and a toroidal pinch with a metal ring at the centre of the plasma held up by magnetic forces, hence its name, the Levitron. The following April he closed another mirror machine, IMP, at Oak Ridge. These terminations sent shock waves through the fusion laboratories. It was the first time that Washington managers, rather than laboratory directors, decided the fate of projects.

Hirsch knew that if fusion was going to be taken seriously by politicians it needed a timetable; an identifiable series of milestones towards a power-producing reactor. He set up a panel of lab directors plus other physicists and engineers to sketch out such a plan. The first milestone they defined was scientific feasibility, showing that fusion reactions could produce as much energy as was pumped into the plasma to heat it – a state known as breakeven. The second would be a demonstration reactor, one that could produce significant amounts of excess energy for extended periods. After that would come commercial prototypes, probably developed in collaboration with industry. The panel suggested that the first goal could be achieved sometime around 1980-82, while a demonstration reactor might be built around 2000. There was some disquiet about this timetable at the fusion labs. They weren't sure eight or ten years was enough to get to scientific feasibility, but at Hirsch's urging this plan became official division policy.

Meanwhile something happened that would give Hirsch's plan new urgency. On 6th October, 1973, Egypt and Syria launched a surprise attack against Israel. Starting on the Jewish holy day of Yom Kippur, the attackers made rapid advances into the Golan Heights and the Sinai Peninsula, although after a week Israeli forces started to push the Arab armies back. The conflict didn't remain a Middle Eastern affair for long. On 9th October the Soviet Union started to supply both Egypt and Syria by air and sea. A few days later, the United States, in part because of the Soviet move and also fearing Israel might resort to nuclear weapons, began an airlift of supplies to Israel. What became known as the Yom Kippur War lasted little more than two weeks, but its effects reverberated around the world for much longer.

Arab members of OPEC, the Organisation of Petroleum Exporting Countries, were furious that the US aided Israel during the conflict. On 17th October, with the war still raging, they announced an oil embargo against countries they considered to be supporting Israel. The effect was dramatic: the price of oil quadrupled by the beginning of 1974, forcing the United States to fix prices and bring in fuel rationing. The prospect of fuel shortages and rising prices at the pump was a shock to the American psyche. All of a sudden the huge gas-guzzling cars of the 1960s seemed recklessly wasteful and US and Japanese carmakers rushed to get more fuel-efficient models onto the market. In response, President Richard Nixon launched Project Independence, a national commitment to energy conservation and the development of alternative sources of energy. That meant more money for fusion, lots more. In 1973 the federal budget for magnetic confinement fusion was $39.7 million; the following year it was boosted to $57.4 million, and that was more than doubled to $118.2 million in 1975. And the increases continued: by the end of the decade magnetic fusion was receiving more than $350 million annually.

Reading the political runes in 1973, Hirsch was keen to get his plan moving but wanted to make one significant change – he wanted the scientific feasibility experiment to use deuterium-tritium fuel. The fusion labs had previously assumed that they would use simple deuterium plasmas so that they wouldn't have to deal with radioactive tritium, radioactive plasma vessels or the added complications of alpha-particle heating. They wanted a nice clean experiment in which they could get a deuterium plasma into a state in which, if it had been D-T, they would get the required energy output – a situation known as 'equivalent break-even.' For Hirsch, that wasn't enough. He suspected that many of the scientists at the fusion labs were just too comfortable working on plasma physics experiments, but he wanted them to get down to the nitty-gritty of solving the engineering issues that a real fusion reactor would face. And he knew that a real burning D-T plasma would be PR gold. The White House, Congress and the public would never understand the significance of equivalent break-even but if a reactor could generate real power – light a lightbulb – using an artificial sun, that would get onto the evening news and every front page.

Not all fusion scientists were against moving quickly to a D-T reactor. Oak Ridge was not afraid of radioactivity. The lab had been set up during the wartime Manhattan Project and had pioneered the separation of fissile isotopes of uranium and plutonium to use in atomic weapons. Since then it had branched out into many fields of technology, some of which involved handling radioactive materials. Oak Ridge's tokamak, Ormak, was performing well and researchers there saw a D-T reactor as a natural next step. In fact, they offered Hirsch more than he had asked for: they proposed a machine that would reach not just break-even but 'ignition,' a state where the heat from alpha particles produced in the reactions is so vigorous that it is enough to keep the reactor running without the help of external heat sources – a self-sustaining plasma. To reach ignition would require very powerful magnets made from superconductors, another area in which Oak Ridge already had expertise.

Seeing all the government money that was being thrown at new energy sources in the winter of 1973-74, Hirsch wanted to speed up his fusion development plan. Instead of building a deuterium-only feasibility experiment by 1980 followed by a D-T reactor by 1987, he proposed that they should move straight to a D-T reactor, starting construction in 1976 and finishing in 1979. Such a timetable would require a much steeper increase in funding as a D-T machine, at $100 million, would be twice the cost of a feasibility experiment.

None of the fusion labs liked this accelerated plan. Princeton didn't want to get involved in D-T burning yet and the new plan would eliminate the deuterium-only feasibility experiment they had hoped to build next. Oak Ridge, although enthusiastic about D-T, thought the timetable was too short. And Los Alamos and Livermore, which were planning new pinches and mirror machines, feared that a big D-T tokamak would consume all the fusion budget and squeeze them out entirely.

An issue that would play a key role in the move to larger tokamaks was plasma heating – how to get the temperature in the reactor up to the level necessary for fusion. Early tokamaks simply relied on ohmic heating, where the resistance of the plasma to the flow of current heats it up. Using ohmic heating alone, these machines were able to get to temperatures of tens of millions of °C, but fusion would need ten times that much. Theory predicted that as the temperature in the plasma got higher, ohmic heating would get less efficient, so another way of heating the plasma was needed. US researchers were pinning their hopes on neutral particle beams. These were being developed at Oak Ridge and the Berkeley National Laboratory as a way of injecting fuel into mirror machines, but tokamak researchers realised that they might work as plasma heating systems.

Neutral beam systems start out with a bunch of hydrogen, deuterium or tritium ions and use electric fields to accelerate them to high speed. If those ions were fired straight into a tokamak they would be deflected because its magnetic field exerts a strong force on moving charged particles. So the ion beam must first be fed through a thin gas where the ions can grab some electrons, neutralise, and move on through the magnetic field undisturbed. Once in the tokamak's plasma, the beam gets ionised again by collisions with the plasma ions but because the beam particles are moving so fast when they do collide they send the plasma ions zinging off at high speed, thereby heating up the plasma.

During 1973 a race developed to see who would be first to demonstrate neutral beam heating in a tokamak. The team running the CLEO tokamak at Culham, using a variation on Oak Ridge's beam system, were first to inject a beam but their measurements didn't show any temperature rise above the ohmic heating. Princeton's ATC, using the Berkeley beam injector, came in next and managed to get a modest rise in temperature. Oak Ridge lost the race, but got the best heating results in Ormak. These first efforts had low beam power, typically 80 kilowatts, and temperature gains were small, around 15% above ohmic heating. But within a year ATC was doing better, boosting the ion temperature in its plasma from around 2 million °C to more than 3 million °C. The signs were promising that neutral beam heating would be able to take tokamaks up to reactor-level temperatures. That would really be put to the test in an upcoming machine, the Princeton Large Torus (PLT), which had begun construction in 1972. Designed to be the first tokamak to carry more than a million amps of plasma current, PLT would have a 2-megawatt beam heating system to boost the ion temperature above 50 million °C.

The sensible thing would have been to wait and see how well the PLT worked before embarking on a larger and more ambitious reactor, but Hirsch didn't want to wait. In December 1973 he called together the laboratory heads and other leading fusion scientists to discuss plans for the D-T reactor. The Princeton researchers were highly critical of Oak Ridge's proposal for a reactor that could reach ignition. Such a machine would require a huge leap in temperature from what was then possible and would cost, they had independently calculated, four times the allotted $100 million budget. Then the question of timetable came up. Oak Ridge's head of fusion Herman Postma was asked if his design would be ready to begin construction in 1976, Hirsch's preferred start date. The reactor's ambitious design and superconducting magnets would take some time to get right, so Postma said that he didn't know. Hirsch was furious.

After a break for lunch, the head of the Princeton lab, Harold Furth, got up and made a surprising proposal. He sketched out a machine that he and a few colleagues had first proposed nearly three years earlier. Since it was difficult to get plasma temperature up to reaction levels, they had reasoned that you could build a tokamak that was only capable of reaching a relatively modest temperature and fill it with a plasma made of just tritium. Then with a powerful neutral beam system they would fire deuterium into the plasma. While there would be no reactions in the bulk of the plasma, at the place where the deuterium beam hits the tritium plasma the energy of the collisions would be enough to cause a reasonable number of fusion reactions. Furth referred to this setup as a 'wet wood burner': wet wood won't burn on its own, but it will if you fire a blowtorch at it.

Such a reactor would never work as a commercial power-producing plant because it could only achieve modest gain (energy out/energy in). Reactor designers had always assumed that beam heating systems would only be used to get the plasma up to burning temperature and then the heat from alpha particles would sustain the reaction. It was never the idea for beams to be an integral part of the reactor. But Furth suggested that this would be a quick and relatively cheap way to get to break-even.

Hirsch gave the two labs six months to come up with more detailed proposals. When those plans were revealed in July 1974, Hirsch had a difficult choice. On one hand was a bold, technologically inventive machine that had the potential to get all the way to ignition in one step, although it came from a lab that was relatively new to the fusion game. The alternative was a much more conservative choice. Princeton's wet wood burner was not dissimilar to the Princeton Large Tokamak that the lab was currently building. It would not go a long way towards demonstrating how a fusion power reactor would work but Princeton had nearly twenty-five years' experience of building these things and if Hirsch simply wanted a demonstration of feasibility, this was more likely to give it to him. He was aware that European labs were working together to design a large reactor and Russia had ambitious plans too, so he could not afford to delay – as always, American prestige was at stake. So he played safe and gave Princeton the nod to begin work on the Tokamak Fusion Test Reactor (TFTR) with an estimated total cost of $228 million.

Like the JET design team across the Atlantic, the designers of TFTR didn't have a lot of information to work with. But unlike Rebut's daring design for JET, the Princeton team opted to keep it simple. No D-shaped plasma for them; they stuck with the tried-and-tested circular design. It was a trademark of the Princeton lab to keep their devices as simple as possible – the simpler they are the faster they can be built and the easier it is to interpret the results.

In 1975 the PLT began operation and produced some impressive results using neutral beam heating, raising the ion temperature to 60 million °C. Theorists had predicted that a beam impacting with the plasma would cause instabilities, but these failed to materialize. Altogether the Princeton researchers were happy with the results, but there was one thing that caused some concern: confinement time got worse the more beam heating was applied. Although this cast a small dark cloud over the future of TFTR, in the rush to finish its design and get construction started there was little time to consider the issue.

Ground was broken for the new machine in October 1977 and it was scheduled to start operating in the summer of 1982. Much of the work was parcelled out to commercial contractors, a significant fraction was built by other government labs, and the rest was done in-house by Princeton staff. Like almost any science project of this size, there were numerous technical headaches along the way. New buildings had to be built with thick concrete shielding to protect people from the neutron flux when D-T reactions were taking place. The power supply system, involving the usual giant flywheel, proved unreliable and had to be virtually rebuilt, which bankrupted the contractor involved. For the first time, computers were bought to help analyse results from the reactor but, being a new technology, it took some years to get them working properly. As the summer of 1982 passed and moved into autumn the diagnostics systems for monitoring the reactor were nowhere near finished. TFTR project director Don Grove was determined for TFTR to get its first plasma before the end of the year, and that meant before Christmas. In desperation Princeton staff took over the installation of diagnostics from the contractor on 12th December and worked around the clock to get it in place.

By 23rd December they had installed the very minimum set of diagnostic instruments. The cabling that connected these in-

**Staff celebrate first plasma in the Tokamak Fusion Test Reactor at Princeton on Christmas Eve 1982. Note the clock, stalled at 1.55 a.m.**

(Courtesy of Princeton Plasma Physics Laboratory)

struments, via a tunnel, to the nearby control room had not been installed so they set up a temporary control room in the reactor building. A thousand and one things had to connected, checked, rechecked and tested. The Princeton researchers had never built such a large and complex machine before and everything was new and unfamiliar. The sky darkened and the team worked on into the evening. Grove decreed that whether they finished or not they would stop working at 2 a.m. The clock ticked past midnight into the early hours of Christmas Eve. They were very close, but not there yet. No one admits to knowing how it happened, but the clock on the control room wall mysteriously stopped at around 1.55 a.m. Since it was not officially 2 a.m. yet, the team kept working.

About an hour later they attempted their first shot. There was a flash in the machine and it was done. Grove ceremoniously handed a computer tape to Furth containing measurement of the first plasma current and Furth handed over a crate of champagne. The giant machine was duly christened and everyone went home for Christmas. TFTR would not produce another plasma till March as the team had to finish installing all the things that had been left out in the rush to meet the deadline.

Despite the head start that the JET team originally had in designing their machine, the delay over deciding its location put them firmly in second place. The ground breaking ceremony for TFTR in October 1977 was only days after the end of the hijack drama at Mogadishu. Construction of JET didn't begin until 1979 so the Culham team were two years behind their US rivals, but they soon made up lost ground thanks to the steely determination of Rebut. The Frenchman did not, however, get the job of JET director. Euratom passed him over and gave the job to Hans-Otto Wüster, a German nuclear physicist. Wüster had made a name for himself as deputy director general of CERN during the construction of its Super Proton Synchrotron. Although not a plasma physicist, Wüster had an easygoing style that allowed him to talk with equal ease to construction workers and theoretical physicists – very different from the blunt Rebut. But underneath the charm he was an adroit politician, something that proved very useful in keeping JET on track. Rebut, however, was knocked sideways by the decision and considered resigning. But the job of technical director still allowed him to supervise the construction of the design he fought so hard to bring to life, and Wüster gave him complete freedom in the construction.

Palumbo also did his bit to set JET off on the right track. Although 80% of JET's funding came from Euratom – with another

10% from the UK and the remaining 10% split between the other associations – he insisted that JET be set up legally under European law as a Joint Undertaking, in other words an autonomous organisation with its own staff of physicists, theorists and engineers, and at arm's length from interference by Euratom and the national labs. Even at Culham, where JET was sited, it remained separate from the national laboratory – it had its own buildings and its own staff. Culham researchers worried that JET would suck all the vitality out of their own lab and tended to view the JET researchers as a superior bunch who kept themselves to themselves. There was one issue that caused more than cool relations: pay. The JET undertaking instituted a system of secondment in which researchers from Euratom association labs would come to JET and work there for a while. During these sojourns they enjoyed the generous rates of pay typical of people working for international organisations. But the Culham staff seconded to JET continued to get the local salary rate which was less than half what their overseas colleagues were getting. This disparity caused huge resentment among British employees at JET, forcing them eventually to take the matter to court.

Just as in the building of TFTR, there were numerous technical hurdles to overcome in JET's construction, but Rebut ruled with an iron hand and allowed very few changes to the design. On 25th June, 1983, almost exactly six months after TFTR produced its first plasma, JET fired up for the first time. 'First light, a bit of current,' the operator wrote in his log book. It was only a bit, just 17 kiloamps, but the race between the two giant machines was on. Rebut had made a bet with his opposite numbers in Princeton that, even though JET started later, it would achieve a plasma current of 1MA first. The loser would have to pay for a dinner at the winner's lab and bring the wine. JET duly passed the milestone first, in October, and so the two teams dined together, at Culham, drinking Californian wine.

* * *

It didn't remain a two-horse race for very long, however, with a Japanese contender known as JT-60 joining in April 1985. Japan had noted the fusion results revealed at the Geneva conference in 1958 but decided against an all-out machine building programme. They kept their plasma physics experiments small and in university labs. When the Russian-inspired dash for tokamaks began in the late 1960s Japan decided to take the plunge and built its first tokamak, the JFT-2, which was roughly the same size as Oak Ridge's Ormak and was completed in 1972. From there, they jumped straight to the giant tokamak class, beginning design work on JT-60 in 1975.

Unlike in Culham and Princeton, the researchers at the new fusion research establishment in Naka did not supervise the construction of the machine themselves. They drew up detailed plans and then handed them over to some of the giants of Japanese engineering, including Hitachi and Toshiba, and left them to get on with it. Although it was a more expensive way of building a fusion reactor, the researchers were spared the stresses and strains of managing a complex engineering project. At Naka there was no unruly rush to demonstrate a half-finished machine just to meet a deadline. Instead the construction companies finished the job in an orderly fashion, tested that it was working, and then handed it over to the researchers. JT-60 was roughly the same size as JET and had a similar D-shaped plasma cross-section. Because of Japanese political sensitivities, it was not equipped to use radioactive tritium so the best it would be able to achieve was an 'equivalent break-even' but despite that, in many ways it exceeded the achievements of its rivals in the West.

Russia too, after the success of T-3 in 1968, continued to innovate, building a string of machines from T-4 right up to T-12. T-7 was the first machine to use superconducting magnets. Superconductors, when cooled to very low temperatures, will carry electrical current with no resistance so they allow much more powerful magnets and much longer pulses. This lets researchers explore how plasma behaves in a near steady state, rather than in a short pulse. The size of T-10 was on a par with Princeton's PLT. Other centres got involved too, such as the Ioffe Institute in Leningrad which built a series of tokamaks.

In the mid 70s, as the US, Europe and Japan began building large tokamaks, Russia too started work on T-15. Although it was not quite as large as the other three giants of that period and was not equipped to use tritium, it was the only one of the four to use superconducting magnets. The intention was to follow T-15 with a dedicated ignition machine, T-20, bigger than TFTR, JET or JT-60. But the researchers' ambitions were undermined by the crumbling state around them. The Soviet Union in the 1980s was already in a downward spiral. The T-15 team had trouble getting funding and materials and the situation got worse every year. By the time the machine was finally finished in 1988 it was already looking out-of-date and the institute couldn't afford to buy the liquid helium needed to cool its superconducting magnets. Following the final collapse of the Soviet Union in 1991 the new Russia just didn't have the resources to run an active fusion programme and T-15 was eventually mothballed.

Back in 1983, the Princeton researchers were getting used to their new machine. The huge scale of the thing compared to earlier tokamaks made it an exciting time. They ran the machine for two shifts each day. Researchers would gather for a planning meeting at 8 a.m. and the first shots would begin at 9 a.m. A roster of thirty-six shots per day was typical. There would be another meeting at 5 p.m. and shots would often run late into the evening but they had to stop at midnight to let the technicians and fire crew go home – they always had to be present for safety reasons. If the

**The Tokamak Fusion Test Reactor at Princeton showing the neutral beam injection system on the left.**

(Courtesy of Princeton Plasma Physics Laboratory)

team were working late they would send one of their number out to get a carload of pizzas or hoagies – the Philadelphia term for a submarine sandwich.

TFTR was bristling with diagnostic instruments and the huge volumes of data these produced demanded a whole new way of working. Smaller machines had essentially been run by one small group of researchers who would plan experiments, carry them out, analyse the results and then do some more. Such an approach on TFTR would waste too much valuable machine time. So different groups were set up with a range of goals so that at any one time some would be preparing experiments, others doing shots and collecting data, and others analysing results of earlier shots. While it was all relatively informal at first, soon there were demands that groups didn't horde their data but made it available for everyone to study. Competition for time on the machine grew so intense that they had to set up a system of written experimental proposals, five to ten pages long, that were peer-reviewed by other researchers at the lab. Princeton had entered the realm of 'big science' and it took some time for its researchers to adjust.

Initially, the researchers were getting very encouraging results. Even though the neutral beam heating systems hadn't been installed yet, TFTR was producing temperatures in the tens of millions of °C using ohmic heating alone and with respectable confinement times. JET, when it started up in June 1983, made similarly good strides using ohmic heating. The scaling laws had been right that larger machines would lead to better confinement. But later, when heating was applied in both machines, the mood changed. TFTR initially had just neutral beam heating while JET had two heating systems, neutral beams and a radiowave-based technique known as ion cyclotron resonant heating or ICRH. In a tokamak plasma the ions and electrons move in spirals around the magnetic field lines and these spirals have a characteristic frequency. If you send into the plasma a beam of radiowaves at the same frequency, the waves resonate with spiralling particles and pump their energy into the particles, boosting their speed and hence the temperature of the plasma. So JET had radiowave antennas in the walls of the vessel to heat the plasma via ICRH.

But whatever the heating method used, the effect was the same: although heating did lead to higher temperatures, as predicted, it produced instabilities in the plasma which led to reduced confinement time. The loss in one counteracted the gain in the other so the overall effect was not much improvement in overall plasma properties. Projections showed that if the plasmas continued to behave in the same way as heating was increased, neither machine would get to break-even. The warnings about neutral

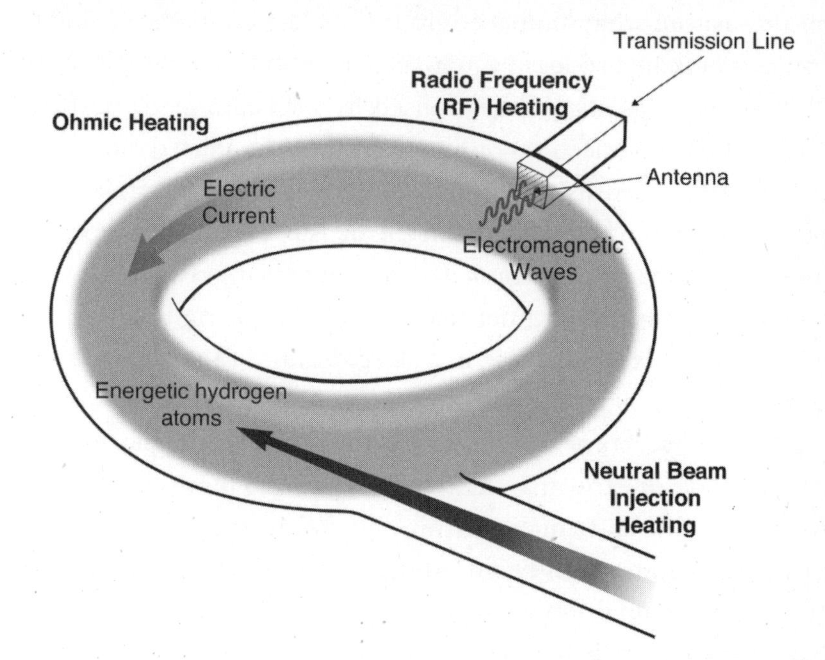


**Tokamaks need help to heat plasma to fusion temperatures, usually provided by ohmic heating (friction), radio waves and neutral particle beams.** (Courtesy of EFDA JET)

beam heating provided by machines such as PLT had come too late: both Princeton and Culham seemed stuck with designs that would not achieve their goals.

In February 1982, Fritz Wagner, a physicist at Germany's fusion lab, the Max Planck Institute for Plasma Physics in Garching near Munich, was carrying out experiments with the lab's ASDEX tokamak. ASDEX was a medium-sized tokamak and Wagner, who was relatively new to plasma physics, was studying the effect of neutral beam injection on the properties of the plasma – this was before TFTR and JET had started up. He started his shots just heating the plasma ohmically and then turned on the neutral beam and measured what happened. In most of his shots the arrival of the neutral beam produced a jump in temperature and the inevitable dip in density as instabilities caused by the beam made particles escape. But he noticed something strange: if he started out with a slightly higher density of particles and stayed above a certain beam power, when the beam kicked in the density suddenly jumped up instead of down and continued to rise, eventually reaching a state where temperature and density remained high right across the width of the plasma, ending in a steep decline at the plasma edge. This was entirely unlike the usual pattern which showed a maximum of temperature and density in the centre of the plasma and a gradual decline towards the edge. Wagner did more experiments with different starting densities and found that there was no in-between state: high beam power was needed and the density either jumped up or down depending on whether the starting density was above or below a critical value.

Wagner was perplexed and spent the weekend checking and rechecking his results to make sure he hadn't misinterpreted something. His colleagues at the Garching lab were sceptical about it at first. No such effect had ever been predicted by theory or seen at another lab. But Wagner was able to demonstrate the effect reliably on demand so they were forced to take it seriously. If the effect worked on other tokamaks it could be amazingly important because the jump up in density, which was soon dubbed high-mode or H-mode, produced confinement twice as good as the low pressure state (low-mode or L-mode). The wider community of fusion scientists took more persuading. At a fusion conference in Baltimore a few months later he was grilled for hours by a disbelieving audience at an evening session. Until some other tokamak could also demonstrate H-mode, it would remain a curious quirk of ASDEX.

Wagner had to wait two years for another tokamak, Princeton's Poloidal Divertor Experiment (PDX), to prove him right.

Another machine, DIII-D at General Atomics in San Diego, repeated the feat in 1986. Now everyone was interested in H-mode. TFTR and JET, which were both struggling with poor confinement brought on by neutral beam injection, could be saved by H-mode but no one knew if it would work in such big machines. And there was another problem: ASDEX, PDX and DIII-D all had something that the giant tokamaks didn't have – a divertor – and it seemed that H-mode only worked if you had one.

A divertor is a device in the plasma vessel that aims to reduce the amount of impurities that get into the plasma. Impurities are a problem in a fusion plasma because they leak energy out and make it harder to get to high temperatures. It works like this: if the impurity is a heavy atom, like a metal that has been knocked out of the vessel wall by a stray plasma ion, it will get ionised by collisions with other ions as soon as it strays into the plasma. But while a deuterium atom is fully ionised in a plasma – it has no more orbiting electrons – a metal atom will lose some of its electrons but hold onto others in lower orbitals. It is these remaining electrons that cause the problem. When the metal ions collides with others these electrons get knocked up into higher orbitals and then drop down again emitting a photon which, immune to magnetic fields, will shoot out of the plasma, taking its energy with it.

In early tokamaks, researchers tried to reduce this effect with a device called a limiter. There were different types of limiter but a common one took the form of a flat metal ring, like a large washer, which fits inside the plasma vessel and effectively reduces its diameter at that point. During operation the plasma current has to squeeze through the slightly narrower constriction formed by the limiter. This helps to reduce the plasma diameter and so keeps it away from the walls, and it also scrapes off the outermost layer of plasma where most impurities are likely to be lurking. As the only place where the plasma deliberately touches a solid surface, limiters had to be made of very heat-resistant metals such as tungsten or molybdenum. But when external heating began to be used in tokamaks in the mid 1970s the higher temperature proved too much for metal limiters and they started to become a source of impurities rather than a solution for them. So researchers switched to limiters made of carbon which is very heat-resistant. Even if the carbon did end up as an impurity it would do less damage because, being a light atom, it would probably be fully ionised by the plasma and wouldn't radiate heat.

Some labs tried to counter the problem of impurities by coating the inside walls of their vessels – usually made of steel – with a thin layer of carbon. The coatings helped but they didn't last for long, so at some labs they began to cover the inside walls of their vessels with tiles of solid carbon or graphite. Russia produced the first fully carbon-lined tokomak, TM-G, in the early 1980s and after it reported encouraging results others followed suit. By 1988 the interiors of JET, DIII-D and JT-60 were half covered in tiles and total coverage only took a few more years.

Limiters were, however, still proving to be a problem and researchers resurrected the idea of a divertor that Lyman Spitzer had first suggested in 1951 for his stellarators. A divertor takes the meeting point between the outer layer of plasma and a solid surface and puts it in a separate chamber, away from the bulk of the plasma, so that any atoms kicked out of the surface could be whisked away before they polluted the plasma. In some of Spitzer's stellarators, at a certain point in one of the straight sections, instead of the narrow aperture of a limiter there would be a deep groove going all the way around the vessel poloidally (the short way around). Extra magnets would coax the outermost magnetic field lines – known as the 'scrape-off layer' – to divert from the plasma vessel and form a loop into the groove and out again. But inside the groove the field lines would pass through a solid barrier so any ions – deuterium or impurity – following those field lines would be diverted into the

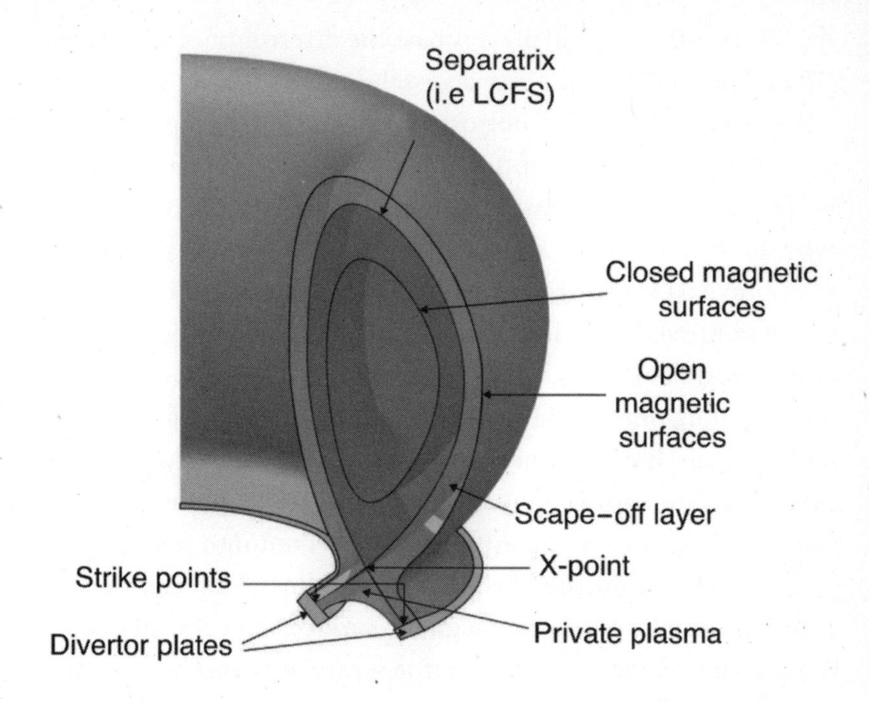


**The divertor at the bottom of a tokamak's plasma vessel removes heat and helium 'exhaust' from the plasma, and helps to achieve H-mode.** (Courtesy of EFDA JET)

groove and then halted by the barrier. Unlike a limiter, this halting of the outermost ions occurs away from the main plasma, where it's less likely to re-pollute it.

Divertors didn't work well in stellarators. The extra fields to divert the scrape-off layer caused such a bump in the magnetic field at that point that it worsened confinement. But in the mid 1970s people tried them again in tokamaks, first in Japan followed by Russia, the UK, the US (PDX) and Germany (ASDEX). In tokamaks it was possible to position the divertors differently: because the plasma moves by spiralling around the plasma vessel – combining toroidal and poloidal motion – a divertor could be fitted as a groove going around the torus the long way, toroidally. In this way the symmetry of the toroidal shape is not spoiled but the scrape-off layer will always pass the divertor once per poloidal circuit. And D-shaped plasma vessels had the perfect place to put a divertor: in the top or bottom corners of the D.

This small group of tokamaks that had divertors seemed to be the only ones in which H-mode worked, but nobody knew why. Both JET and TFTR were desperate to try to reach H-mode to improve their performance, but they were designed before divertors had proved themselves in tokamaks so neither had one. JET at least had the D-shape that could easily accommodate a divertor but installing one would involve an expensive refit. The JET team had a hunch, however: perhaps it was not the divertor itself that was responsible for H-mode but the unusual magnetic configuration with the outer scrape-off layer pulled out into a loop.

In the bulk of the plasma, the field lines loop right around forming closed, concentric magnetic surfaces, like the layers of an onion. The magnetic surfaces in the scrape-off layer are said to be open surfaces because they don't close the loop around the plasma but veer off into the divertor. There is one surface that marks the boundary between the open and closed magnetic surfaces. Known as the 'separatrix,' this surface appears to form a cross – dubbed the x-point – close to the divertor where field lines cross over themselves.

In H-mode, the plasma edge was marked by a steep drop in density and temperature, almost as if something was blocking plasma from escaping. JET researchers wondered whether this 'transport barrier' was in some way related to the separatrix, the transition from closed to open magnetic surfaces. The question for JET was whether it could reproduce this magnetic shape with a separatrix and x-point without having a divertor? And if they could, would it produce H-mode? By adjusting the strength of certain key magnets around the tokamak, JET researchers were able to stretch out the plasma vertically and, eventually, produce the desired shape. The x-point was *just* inside the plasma vessel and the open magnetic field lines simply passed through the wall instead of into a divertor. It was enough to have an attempt at H-mode. They tested JET in this divertor-like mode for the first time in 1986. With a plasma current of 3 MA and heating of 5 MW, the plasma went into H-mode for 2 seconds, reaching a temperature of nearly 80 million °C and holding a high density. Researchers calculated that if they had been using a 50:50 mix of deuterium and tritium they would have produced 1 MW of fusion power. So H-mode was possible in a large tokamak. All JET needed now was a divertor.

TFTR, however, was stymied. With its circular vessel cross-section it was difficult to install a divertor and almost impossible to coax its magnetic field into an elongated shape with x-points and open field lines. So instead the Princeton researchers chipped away at the problem in any way they could think of, trying to coax longer confinement times despite the degradation caused by neutral beam heating – and eventually they did make a breakthrough, almost by accident.

A gruff experimentalist called Jim Strachan was doing some routine experiments on TFTR in 1986, trying to produce a heated plasma with very low density. The problem was the tokamak was not playing ball. Ever since they had started coating parts of the vessel interior with carbon – to stop metal from the walls from getting into the plasma – they had encountered a downside of carbon: it likes to absorb things. Carbon will absorb water, oxygen and hydrogen, along with its siblings deuterium and tritium. With all this stuff absorbed into the vessel walls, when you start heating a plasma the heat causes the absorbed atoms to emerge again and contaminate the plasma. Even if it is just deuterium in the walls – the same stuff as the plasma – it means that experimenters had no control over the plasma density because they never knew how much material would emerge from the carbon.

Earlier in 1986 a new carbon limiter had been installed in the TFTR vessel. This wasn't a narrowing ring at one point in the torus but was instead a sort of 'bumper' of carbon tiles along the midline of the outer wall right around the torus. Once installed, the limiter was saturated with oxygen so researchers ran hot deuterium plasmas in the tokamak to oust the oxygen. That worked fine but it left the carbon tiles full of deuterium which would play havoc with Strachan's attempt to produce low density plasmas. So Strachan started running shot after shot of helium plasma to get rid of the deuterium. Helium is a non-reactive noble gas, so does not get absorbed into the carbon as much. Strachan continued this for days, trying to get the tokamak as clean as possible, and then on 12th June he did a low-density heated deuterium shot – TFTR's shot number 2204. The density stayed low, the temperature high and, astoundingly, the confinement time – 4.1 seconds – was twice what TFTR had achieved before. What's more, it produced lots of neutrons – a sign of fusion reactions.

Strachan tried again and found he could produce these shots – which his colleagues soon dubbed 'supershots' – on demand. The key seemed to be the preparatory cleaning shots using helium, so the TFTR team set up a protocol to prepare the machine that way before every shot. This conditioning took anywhere between two and sixteen hours for every supershot, but it was worth it. The plasma current had been low in Strachan's initial efforts, but with further experiments researchers managed to push the plasma current up, until at last they were working with plasmas that were a lot more like the ones needed for a fusion reactor, with temperatures above 200 million °C. Finally, the TFTR team could again see a path to D-T shots and alpha-heating.

JT-60 was the only one of the three giant reactors to have been built with a divertor, but it was positioned half way up the outside wall. When the Naka team tried to produce H-mode, they found that they just couldn't get the right magnetic configuration with the divertor where it was. The Japanese researchers worked with the reactor for just four years and then in 1989 made a bold decision: they gave JT-60 a complete refit, installing a new divertor at the bottom and covering the whole interior with carbon tiles. The rebooted JT-60U began operating again in 1991 and the gamble paid off because it was soon operating in H-mode with properties as good as JET's.

By the beginning of the 1990s, TFTR and JET had refined supershots and H-mode to such an extent that they were getting fantastic results. TFTR could reach ion temperatures of 400 million °C. One measure of success is gain, the ratio of fusion power out over heating power in, denoted by *Q*. So break-even would be *Q*=1. At a conference in Washington in 1990, the two teams reported shots that, if they had been performed with D-T plasma, would have achieved *Q*=0.3 for TFTR and *Q*=0.8 for JET. Two years later the JET team announced that they had produced shots that would be *Q*=1.14 in D-T – more power out than in – but this record was soon bettered by JT-60U, which achieved *Q*=1.2. Although it had taken longer than expected for the big tokamaks to achieve this sort of performance, because of the problems with degraded confinement, they had got there. But just as the teams at Princeton and Culham were starting to think about tritium and burning plasmas something unbelievable happened: a pair of scientists declared that they had achieved fusion in a test tube.

On 23rd March, 1989, Martin Fleischmann, a prominent electrochemist from Southampton University, and Stanley Pons of the University of Utah stood up in a press conference in Utah and described an experiment they had been performing in the basement of the university's chemistry department. They took a glass cell – little more than a glorified test tube – and filled it with heavy water – made from deuterium and oxygen. They inserted two electrodes, one of platinum and one of palladium, and then they passed an electric current through it. Nothing much would happen in the cells for hours or even days but then they would start to generate heat; much more heat, Fleischmann and Pons said, than can be explained by the current passing through the cell or any chemical reactions that might be taking place. Their best cell, Pons said, produced 4.5 watts of heat from 1 watt of electrical input: $Q$=4.5. They didn't believe that a chemical reaction could be producing such heat and so it had to be a nuclear process, in other words the fusion of deuterium nuclei into helium-3 and a neutron.

Part of what was going on in the cells was an everyday process called electrolysis. When they passed a current through the cell, heavy water molecules were split apart and the oxygen ions migrated towards the positive, platinum electrode while the deuterium ions moved to the negative, palladium electrode. But the choice of electrodes was key because palladium is well known to have an affinity for hydrogen, and hence for deuterium too. It is able to absorb large quantities of hydrogen or deuterium into its crystal lattice structure. During the hours when the cells are first switched on, the palladium electrodes absorb more and more deuterium. Pons believed that eventually there would be twice as many deuterium ions in the lattice as there were palladium ones.

The next part is where it all becomes strange. Pons and Fleischmann believed that some of these deuterium ions, crushed together in the palladium lattice, somehow overcame their mutual repulsion and fused. As evidence for fusion, Fleischmann and Pons said they had detected neutrons coming from the cells – which would be expected from such a fusion reaction – as well as both helium-3 and tritium – other possible fusion products. The two scientists were cautiously optimistic about the usefulness of their discovery. 'Our indications are that the discovery will be relatively easy to make into a usable technology for generating heat and power,' Fleischmann told the press conference. The Utah announcement caused a sensation around the world. Newspapers, TV and radio picked up the story. The idea that you could generate as much heat and electricity as you wanted using a simple glass cell filled with deuterium from seawater fired people's imaginations. No longer would the world be dependent on coal, oil, natural gas and uranium.

The initial reaction of fusion scientists was utter incredulity. It just didn't make sense. As far as they had always understood it, deuterium ions are extremely reluctant to fuse because of their positive charges. Those charges make the ions strongly repel each other and it takes a huge amount of energy to force them close enough together to fuse. Inside a metal lattice, where was all the energy coming from to overcome the repulsion? But there were reasons for fusion scientists to pause for thought. The behaviour of ions inside a metal lattice is much stranger and harder to predict than in the near empty space inside a tokamak. Lattices do funny things to ions, affecting their apparent masses and how they interact with other ions. This was an environment that most plasma physicists had little knowledge of. Could it be that it was just something that they had missed? The two scientists were highly respected – Fleischmann was one of the world's foremost electrochemists. And these two men must have been pretty sure of their results to stand up and make such bold claims in front of the world's press, without the usual procedure of having published their results in a journal first and subjecting them to review by other experts.

In the days that followed the press conference, it also emerged that another group at a different Utah institution – Brigham Young University – had been doing similar experiments and had also detected neutrons and heat, but much less heat than Fleischmann and Pons found. Also fusion researchers found themselves called upon in the media to explain what fusion is all about, and to justify why they needed such huge complex tokamaks to achieve it. All of a sudden these machines seemed a wasteful extravagance.

One of the benefits of such a simple experimental setup as the one in Utah is that it is very easy for other scientists to carry out similar tests to verify or refute it. Within days researchers the world over had current flowing through heavy water and were waiting to see the same signs of fusion taking place. It took a few weeks for the first results to come in. A team at Texas A&M University also detected excess heat in their cells, which were modelled on Pons and Fleischmann's, but they hadn't yet tested for neutrons. Other results came in from labs all over the world in the following weeks, but they didn't make things any clearer: some saw neutrons but not heat, others got heat but no neutrons, and some found nothing at all. Researchers at Georgia Tech announced they had found neutrons, only to withdraw the claim three days later when they realised their neutron detector gave false positives in response to heat.

That didn't stop Pons being greeted as a hero when he appeared at a meeting of the American Chemical Society in Dallas on 12th April. Chemists were enjoying their moment in the Sun. There was a recent precedent of a miraculous discovery by a pair of researchers working on their own. Three years previously physicists had been stunned when two researchers in a lab in Switzerland announced the discovery of high-temperature superconductivity. That led to a now legendary session at the March 1987 meeting of the American Physical Society in New York City. Thousands of scientists crammed into a hastily arranged session – now referred to as the 'Woodstock of physics' – to hear all the latest results about these wondrous new materials. For the 7,000 chemists who gathered in Dallas to hear about cold fusion, it was their turn to make history.

The president of the Chemical Society, Clayton Callis, introduced the session by saying what a boon fusion would be to society and commiserated with physicists over what a hard time they were having achieving it. 'Now it appears that chemists have come to the rescue,' he said, to rapturous applause. Harold Furth, director of the Princeton fusion lab, came to make the case for conventional fusion. He didn't think nuclear reactions were happening in the Utah pair's cells. Certain key measurements hadn't been done so the proof for fusion just wasn't there. But that wasn't what the chemists wanted to hear. During his presentation, Furth had shown a slide of his lab's giant tokamak, TFTR – the size of a house, bristling with diagnostic instruments and sprouting pipes and wires. When Pons came on afterwards he flashed up a slide of his own setup: a glass cell the size of a beer bottle held by a rusty lab clamp in a plastic washing-up bowl. 'This is the U-1 Utah tokamak,' he said, and the crowd went wild.

Despite the enthusiasm of the chemists, people were starting to ask hard questions about cold fusion. One was over the shortage of neutrons. Those labs that did see them never found very many. If the cells in Fleischmann and Pons' experiments were really producing as much heat as they claimed via conventional fusion reactions, they should be producing enough neutrons to kill the experimenters. In fact researchers were detecting one-billionth the number that would be expected. Some suggested that this was a new form of fusion that produces heat but no neutrons. More results came in from other labs, but they were still contradictory. One troubling observation was the erratic nature of the heat produced: sometimes cells would run for days producing nothing, then put out a burst of heat, continue for a while then stop.

Fleischmann and Pons didn't help the situation by being reluctant to give out too much information about their experiment. The paper that they submitted to *Nature* just after the March press conference was returned with questions from the reviewers

– a standard procedure – but Fleischmann withdrew it from publication saying he was too busy to make the asked-for revisions. In public, their responses to questions were often evasive and sometimes downright cryptic. At the Dallas chemistry conference Pons was asked why he hadn't done a control experiment using normal water. Although normal water and heavy water have different nuclei, they are chemically identical so if the effect they were seeing was a chemical reaction then a cell filled with normal water should behave in exactly the same way. Such a control experiment would be an obvious thing for any experienced scientist to do. But Pons said using normal water was not a good control. Asked why not, Pons suggested that they had done it and had seen fusion. 'We do not get the total blank experiment we expected,' he said.

On 1st May the American Physical Society had its meeting in Baltimore. It also scheduled a special session on cold fusion but this was no Woodstock. By this time, and with this audience, the mood was very different. Neither Fleischmann nor Pons attended the Baltimore meeting but a seventeen-strong team of chemists and physicists from the California Institute of Technology described their attempts to replicate the Utah experiments. They found many places where mistakes could have been made and concluded that you could explain the results without resorting to fusion. 'We're suffering from the incompetence and delusions of Professors Pons and Fleischmann,' said Caltech theoretical physicist Steven Koonin. 'The experiment is just wrong.' At the end of the session the nine main speakers were polled and eight declared they thought cold fusion was dead, the ninth abstained.

Meanwhile, Fleischmann and Pons were in Washington demonstrating their apparatus to members of Congress while officials from the University of Utah tried to persuade the politicians to provide as much as $40 million for a $100-million cold fusion research centre in Utah. Congress declined but the Department of Energy was asked to investigate cold fusion. The DoE report, published in July, said that it doubted the Utah results were the signs of a new nuclear process and, in any event, whatever was going on in the cells wasn't going to provide a useful source of energy. So there would be no federal cold fusion research programme, but the report said that enough questions remained for the Department of Energy to fund a few studies through normal channels.

By now, only a few months since it was born, cold fusion was running out of friends. The major US newspapers had long since stopped running stories. Most labs had quietly given up on their cold fusion studies but there remained a determined few groups who were convinced that, whether or not it was fusion, there was something interesting going on in these cells that deserved investigating. The state of Utah agreed and put up $4.5 million in August to set up the National Cold Fusion Institute. But for most of the scientific world, the cold fusion saga was over. Many now consider it an example of 'pathological science,' where a combination of wishful thinking and misinterpretation of complex results causes researchers to hold fiercely to an idea that most have dismissed. Fleischmann and Pons left the US in January 1991 and set up a lab in France funded by the Toyota car company. The lab was closed in 1998 having still not achieved fusion in a bottle. The National Cold Fusion Institute closed in June 1991 when it ran out of money.

Although the whirlwind of excitement around cold fusion did not last long, it cast an uncomfortable spotlight on the state of conventional, hot fusion. Scientists had been working on it for more than four decades and had spent billions on increasingly complex machines, but still they had generated no excess heat. But by the early 1990s, most felt that they were close. The big tokamaks had shown, using deuterium alone, they could get plasmas dense enough and hot enough that – if there had been tritium in there – a large number of fusion reactions would take place and the alpha particles generated would start to heat the plasma themselves. The researchers at Princeton and Culham, however, were in no hurry to move to D-T operation. They were learning a lot with their machines about how to get good confinement, how to create long stable pulses, and how to control instabilities. Moving to D-T would make everything more difficult: the reactors would need more shielding to protect people from neutrons; every scrap of radioactive tritium would have to be accounted for; and the bombardment of neutrons would make the reactor itself moderately radioactive – or 'activated' – so any later modifications inside the vessel would be much more complicated.

The Culham researchers were coming under some pressure from the JET council to move ahead with D-T experiments. The reactor was working well; the tritium handling facilities were built; the council wanted to see some energy. But Rebut and his team had another good reason to delay D-T: they wanted to refit the interior of JET with a divertor, which was now thought to be essential for a future power-producing reactor. Apart from the role it plays in H-mode, a divertor would be able to extract the fusion exhaust – the alpha particles – and stop the plasma getting clogged up with them. Getting experience using a divertor would pay dividends when planning future reactors. Here was the dilemma: installing the divertor first would delay D-T experiments for too long; doing D-T first would make the interior activated so the whole refit would have to be done with remote handling, a slow and laborious process. So they came up with a compromise plan: they would do some shots with a plasma of 90% deuterium and 10% tritium. This would provide real evidence of plasma burning, but would keep the rate of neutron production at a low level so the vessel interior wouldn't get too activated.

**The interior of JET, showing limiters on the central column, radio antennas on the outside wall and a divertor at the bottom.** (Courtesy of EFDA JET)

There was much to be done to get the reactor ready for D-T operation. The shielding had to be checked, the tritium handling tested, two of JET's sixteen neutral beam sources had to be set up to fire tritium instead of deuterium, and the researchers had to work out what was the best kind of pulse to get the plasma to burn. They settled on a plasma current of 3 MA, a toroidal magnetic field of 2.8 tesla and 14 MW of neutral beam heating – experiments with deuterium showed that as soon as the heating beams kicked in this plasma swiftly went into H-mode and the number of neutrons produced from D-D fusions continued to grow. It was a time of high excitement: many of those who worked on JET had spent their entire working lives trying to make a plasma that would burn; now they were going to see if it was all worthwhile. Media organisations got a whiff that something was going on and asked if they could be present at the first burning plasma. Rebut thought about it and decided that the public had paid for it and so the public had a right to know what was going on. It was a potentially risky strategy, however, because it could fail to work. Fusion researchers had been burned before with ZETA and didn't want the same thing to happen again.

On the appointed day, 9th November, 1991, hundreds of people crowded into the JET control room – researchers, officials, journalists. Getting the machine ready wasn't a quick process. First the researchers did some shots with just deuterium to check they were still getting the plasma performance they wanted. Then they carried out some shots containing a trace of tritium, less than 1%, to test the diagnostics. Then the moment of truth arrived. The crowds in the control room craned to see screens that showed an image of what was going on inside the vessel. As the shot started they could see the diaphanous plasma form inside the vessel and when controllers switched on the neutral beams, including those firing tritium, the screens whited-out as neutrons flooded the camera. The control room erupted into applause. They had achieved the first controlled release of significant amounts of fusion energy.

The power had peaked for only around two seconds and reached a maximum value of 1.7 MW which amounts to Q=0.15, although this would have been Q=0.5 if a 50:50 D-T mixture had been used. JET was lauded in headlines and news bulletins around the world. Although the Culham researchers only performed two D-T shots, they had entered the era of burning plasma and they had done it before Princeton. But it would be a while before they would be able to do it again because JET was soon shut down to fit a divertor and the spotlight moved across the Atlantic.

* * *

JET's success caused some gnashing of teeth in Princeton because researchers there felt they could have got their first. Bob Hunter, their boss at the Department of Energy at the time, wanted a better understanding of the plasma before moving on to D-T so they continued to perfect the art of performing supershots. They consoled themselves with the thought that 10% tritium wasn't 'proper' fusion, and that they would be first to produce a fully burning plasma.

Just as at Culham, there was an enormous amount of work to do to get ready. It was all hands on deck with large crews of physicists and engineers working to prepare TFTR, doing double shifts and working on Saturdays. Because the lab was close to the town of Princeton and they would be bringing radioactive tritium onto the site, lab director Harold Furth worked hard to reassure local residents. The scientists held open meetings, explained their plans and answered questions. Their tritium handling facilities were reviewed, and reviewed again. At one stage a 'tiger team' of as many as fifty people descended on the lab for a week to thoroughly inspect all their procedures.

Finally the day arrived. Much thought had gone into how to publicise the event. Journalists from *The New York Times* and other publications were invited and some spent most of the week there. The lab employed its own film crew to record events and then to produce short items for TV stations to use. There were so many people there that most had to watch events on screens in the lab's auditorium. Everyone was given identity badges with colours signifying their level of access – only those with the coveted red badge were allowed into the TFTR control room. After each preparatory shot was made, researchers would come to the auditorium to update the audience on what was happening. The team was deploying its full arsenal of neutral particle beams, with a total power of nearly 40 MW. The control room had an area called 'beam alley' where twelve people sat each tweaking one of the dozen beam sources to get maximum power out. Instead of a camera looking into the vessel, the TFTR team had positioned a sheet of material called a scintillator inside the neutron shielding next to the reactor and a camera was set up looking at the scintillator. When a neutron hit the scintillator it would produce a small flash of light so the amount of flashes indicated the rate of neutron production. The audience watched as preparatory shots produced a few specks of light on the screen. When it finally came time for the 50:50 D-T shot, the scintillator became a bright glowing square and cheers filled the auditorium. They had produced 4.3 MW of fusion power. It was more than four decades since Lyman Spitzer, there in Princeton, had dreamt of building a fusion reactor and started the project that became PPPL. The researchers there that day genuinely felt they had made history.

TFTR did not, like JET, do two shots and then close down. Instead it embarked on an extended programme to study D-T fusion. Furth was determined to get to break-even and also set a first interim goal of 10 MW of fusion power which his staff went after tenaciously. The following year they were closing in, having got powers of more than 9 MW. But controllers of the machine were getting nervous because everything was being pushed to the limit: conditioning the vessel was exhaustive, they were using the highest possible magnetic field and plasma current, and the neutral beams were turned up to maximum. Then, under these conditions, they did a shot that ended in a huge disruption – this is where there is a sudden loss of containment and all the energy is dumped in the structure of the machine. There were microphones set up in the reactor hall so that they could hear what was going on over in the control room, which was in a separate building. The team heard a noise that sounded like the hammer of hell, followed by echoing thunder and then absolute quiet. The whole reactor building had been shaken. There was no serious damage, but after that they were much more careful. They tweaked the plasma to make it less prone to disruptions and later that year reached a power of 10.7 MW.

During four years of experiments the Princeton researchers learned a huge amount about controlling a burning D-T plasma and set many more records. They were the first to demonstrate self-heating of the plasma by alpha particles. They set a record temperature of 510 million °C, a record plasma density of 6 atmospheres, among many other things. What TFTR didn't achieve was break-even, only reaching a gain of Q=0.3. TFTR simply wasn't big enough for that kind of performance: a larger plasma insulates the ions in the centre from the outside so they have more time to react, and TFTR's traditional circular cross-section didn't allow it to get the same sort of plasma currents that D-shaped reactors could.

Then, in April 1997, TFTR was closed down. Many at the lab thought there was more that could have been done with the reactor. There were plans for further upgrades, but TFTR was an expensive machine to run, with its large staff and its tritium facilities, and the Department of Energy didn't have the money to keep it going. What disturbed fusion researchers more was the fact that there were no plans for a replacement. PPPL was never normally slow to begin planning for the next machine. Back in 1983, when TFTR had only just started operating, researchers there were already working on the design of its successor which would study in detail the physics of burning plasmas heated by alpha particles and go all the way to ignition. But as the design of this Compact Ignition Tokamak progressed and its designers learned more from operating TFTR, it grew larger in size and cost. Then a review of the design ordered by the Department of Energy cast doubt on CIT's ability to reach ignition and so the design was scrapped in 1990. Princeton responded with another plan, the Burning Plasma Experiment, which sought to address the deficiencies of CIT. But BPX also grew bigger and even more expensive than CIT and by this time the US was participating in the design of ITER, the international fusion reactor project, that would do everything BPX could do and more. So BPX was also abandoned in 1991. Princeton bounced back again with the Tokamak Physics Experiment, a smaller-scale machine that aimed to complement ITER by studying such issues as steady-state operation.

This TPX, being small, was more in tune with the times. Since the golden era of the late 1970s and early 1980s, when Middle East oil embargoes had pushed the government to spend huge amounts researching possible alternative sources of energy, funding of fusion research had been following a steady downward trajectory. By the early 1990s, the DoE's fusion budget was around half what it had been at its peak in 1977. Fusion scientists had had to temper their ambitions over the years, but worse was yet to come. The autumn of 1994 saw the 'Republican revolution' when the Republican Party won control of both houses of Congress for the first time in more than forty years. As they cast around looking for superfluous government spending to cut, fusion's track record did not look good. Over forty years taxpayers had spent more than $10 billion on fusion and all those expensive machines had not even reached break-even.

In the Republicans' first budget, for 1996, fusion saw its roughly $350 million annual funding cut by more than $100 million. Something had to go and that meant first TPX and then TFTR itself. Some Princeton staff took the TPX design to South Korea where they helped the country build a scaled-down version as the Korea Superconducting Tokamak Advanced Research (KSTAR) facility, which produced its first plasma in 2008. But US researchers were left after 1997 with no major machine to work on. The DoE changed the emphasis of its fusion programme from one aimed at fusion energy to one investigating the science required to achieve fusion. Its main project was now ITER, but even that didn't prove to be an easy relationship.

* * *

In the same year that TFTR closed, JET researchers were agitating to have another crack at D-T operation but were being held back by the Euratom council. In the six years that had passed since their two 10% tritium shots they had fitted JET with a divertor and had experimented with it using deuterium-only plasmas. From the council's point of view JET was proving to be a valuable research tool for learning how to control plasmas in a divertor configuration, able to test different strategies and materials. The plan was to fit another more advanced divertor but doing D-T experiments first would make such modifications more difficult. The council also feared expensive decommissioning costs when JET ended its working life if the vessel became very activated. The person in charge was now German plasma physicist Martin Keilhacker, Rebut having left in 1992 to lead the ITER design team. He supported the researchers' view that really seeing what JET could do with D-T shots would provide invaluable data for ITER or any future burning plasma machine. And, he argued, JET's remote handling system would soon be fully operational. The system's many-jointed arm was capable of unfastening, removing and replacing components inside the vessel so the divertor could be upgraded without a person going inside.

Keilhacker won the council round and in September 1997 JET broke TFTR's record with a shot producing 16 MW of fusion power – a gain of $Q$=0.67. There was much less media hoopla surrounding this breakthrough compared to the shots in 1991 but it did make the newspapers. What didn't get any column inches was another experiment performed on JET which was potentially much more important for future fusion energy production. The shots that achieved the highest power never lasted very long. That was because H-mode only works as long as the plasma density doesn't go above a certain critical value. With the neutral particle beams pumping in more particles and H-mode keeping them trapped with its high confinement, the plasma density can only go up and after a couple of seconds H-mode breaks down and the shot terminates. Creating fusion with a succession of short pulses is not ideal for a power-producing reactor; much better would be one that can operate in a steady, unchanging fashion. Achieving that with H-mode would require a mechanism for leaking out some ions so that the density doesn't go past the critical level. The JET researchers attempted to do this by exploiting a new plasma instability peculiar to H-mode known as edge-localised modes, or ELMs.

ELMs are eruptions of plasma that allow fusion fuel to burst out of the plasma's edge towards the wall of the reaction chamber. Because the confinement provided by H-mode is so good, pressure can build up in the plasma and ELMs are the plasma's way of letting off steam. The eruptions come in all sizes but the larger ones are potentially damaging to the reactor as the ejected plasma can hit the vessel wall. Even if that doesn't happen, plasma bursting through the separatrix will get swept by the open field lines down into the divertor and if it is a big burst the divertor can be damaged by the large amount of hot plasma.

JET's peak power shots were done in a way that suppresses ELMs, so that they got maximum confinement. But if you want your shot to last longer you can encourage small ELMs to occur so that some ions leak out and so keep the plasma density down. The JET researchers tried this, a so-called steady-state ELMy H-mode, and were able to produce pulses with a power of 4 MW but which lasted for five seconds. Although this configuration sacrificed some confinement and hence had a lower peak power, it provided a glimpse of a steady-state mode of operation that will likely be used on ITER.

Some might view the big tokamaks as failures. They didn't achieve their main goal of break-even – except, perhaps, JT-60 which reached an equivalent gain of $Q=1.2$ – and it took roughly a decade longer for them to reach their peaks than had originally been planned. Viewed another way, they presided over an era of huge progress in fusion science. They were designed only a few years after tokamaks were widely adopted around the world and their designers knew very little about what made tokamaks tick or how a large one would behave. Once they started and operators discovered how heating degraded confinement, their prospects looked very bleak. But through a mixture of luck and ingenuity they developed modes of operation – supershots and H-mode – that overcame that problem. While TFTR and JET didn't reach break-even they got close enough to know that it was possible – with a bit more current or a slightly stronger field they might have made it.

What the big tokamaks did show was that it was *scientifically feasible* for a controlled fusion reaction to produce more energy than it consumes, a goal that thousands of scientists had been pursuing ever since the likes of Peter Thonemann, Lyman Spitzer, Oleg Lavrentyev and others first dreamed of building a fusion reactor. And the machines did more than that: they showed that alpha particles will heat the plasma, which will be essential for creating a viable power reactor; and they demonstrated modes of operation, such as JET's ELMy H-mode, that will be required for future reactors to run stably and safely for years on end.

Euratom had planned to close JET in 1999 to free up resources for the construction of ITER. But ITER was not about to start construction; in fact the collaboration that designed it was on the point of breaking apart. With fusion's great hope in a critical condition it seemed foolish to close down the next biggest tokamak, and so JET won a reprieve. Management of JET was handed over to the UK Atomic Energy Authority and the machine became a facility that fusion researchers from across Europe could come and use, mostly testing techniques that would be used on ITER when the giant was returned to health.

CHAPTER 6

# Fusion by Laser

THE FIRST SUSTAINED FUSION REACTION ON EARTH PRODUCing excess energy took place on 1st November, 1952, at precisely 7.15 a.m. The venue was the island of Elugelab in the Enewetak atoll of the South Pacific's Marshall Islands. Nearly 12,000 people, military and civilian, were involved in setting up the test, which had the codename 'Ivy Mike.' They built a cryogenics plant on one of the other islands in the atoll to produce liquid deuterium for the fusion fuel. They built a 2.7-kilometre causeway linking four of the islands. For the device itself they built a large corrugated iron building. It needed to be big because the device, nicknamed 'the Sausage,' was more than 6m tall and 2m in diameter. Along with all the cryogenic equipment to keep the liquid deuterium cold, it weighed 74 tonnes.

When the device was detonated, the blast was equivalent to more than 10 million tonnes of TNT, 450 times the power of the bomb dropped on Nagasaki in 1945. A mushroom cloud rose 37 kilometres into the air and spread out across 160 kilometres. Radioactive coral debris fell on ships moored 50 kilometres away. An hour later, after the mushroom cloud and steam had dispersed, a helicopter flew over the site. The islands of the atoll were stripped clean of vegetation except for Elugelab, of which nothing remained.

Five thousand miles away in California Edward Teller, who had been the driving force behind the United States' H-bomb programme, knew that Ivy Mike had been a success without having to wait for the phone call. A quarter of an hour before the scheduled time of the test he had walked across the grounds of the University of California, Berkeley, to Haviland Hall and sat down by the seismometer in its basement. The machine shone a fine point of light onto a photographic plate to record any movements of the Earth. Teller sat in darkness watching the bright spot and, at precisely the time he had predicted it would happen, the light started to dance around wildly on the plate as the Earth shook in response to the blast on the far side of the Pacific. This completed his ten-year-long campaign to ensure that the United States possessed the most devastating weapon imaginable before its enemies did. Teller sent a telegram to his colleagues at the Los Alamos laboratory containing just the pre-arranged phrase: 'It's a boy.'

In 1941 American physicists first began talking about the possibility of constructing a nuclear fission weapon, or A-bomb, amid the concern that Germany may already be well on the way to constructing one. Then Enrico Fermi casually mentioned to Teller that perhaps it would be possible to use an A-bomb to ignite a more powerful fusion weapon. Teller became obsessed with the idea. Soon both men were enrolled into the Manhattan Project and at its newly built headquarters at Los Alamos, Teller argued that the project should also try to build a fusion bomb, or 'Super,' at the same time. His colleagues were not persuaded and the A-bomb remained the focus, but Teller often ignored the work assigned to him so that he could continue to work on the Super. Ironically, the person who ended up taking over Teller's unfinished work was Klaus Fuchs, who was later unmasked as a Soviet spy.

After bombs had been dropped on Hiroshima and Nagasaki,

Teller was all for moving straight on to develop the Super. He feared that the Soviet Union would soon catch up with the US in nuclear weapons technology and he wanted his adopted country to hold onto the advantage it had. But few others were interested. Many participants in the Manhattan Project, having seen the destruction their creation had wreaked in Japan, were appalled at the idea of building a weapon a thousand times more powerful. Many argued that the only feasible use for such a weapon would be to kill huge numbers of civilians; hence it was a weapon of genocide. Some simply thought such a bomb wouldn't work. Most veterans of the project returned to their universities after the war ended.

Then came the first Soviet A-bomb in 1949 and President Truman's crash programme to develop the Super as quickly as possible. The problem was, Teller's design didn't work. He had assumed that the intense heat created by an A-bomb as it ignited would be enough to spark fusion in nearby deuterium-tritium fusion fuel. He and his colleagues at Los Alamos tried out various designs that attempted to get the fusion fuel as close as possible to the exploding A-bomb, such as a spherical A-bomb surrounded by a layer of D-T or the reverse, a sphere of D-T surrounded by a hollowed-out fission bomb. But according to their experiments and calculations, this was not enough to ignite a fusion burn.

The breakthrough came in 1951 when Stanisław Ulam devised some changes that allowed Teller to move forward and, a year later, Elugelab atol was blown out of the water. The design of the Sausage and the more compact H-bombs that came after it remains a military secret but a rough picture of what came to be known as the Teller-Ulam design has been pieced together from a number of sources, including declassified documents and unintentional leaks by former weapons designers. The key innovations of the Teller-Ulam design are that the A-bomb 'primary' should be separated from the 'secondary,' which is a mixture of fission and fusion

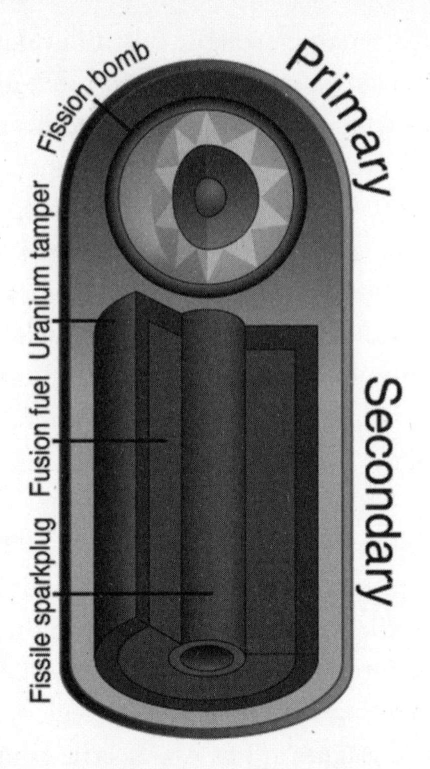


**In the Teller-Ulam design, x-rays from a fission A-bomb primary are used to compress the H-bomb secondary, sparking fusion.** (Courtesy of Wikimedia Commons)

device; and that it is the x-rays generated by the primary, not the heat, which causes the secondary to ignite.

The Sausage was that shape because it had a conventional A-bomb at one end and the hybrid secondary at the other. The multi-stage process of detonation must happen extremely quickly before the blast of the A-bomb blows everything apart. It goes something like this: The A-bomb is detonated and within 1 millionth of a second it is more than three times the temperature of the Sun's core and emitting most of its energy as x-rays. The tubular container of the device is a radiation case, known as a hohlraum, made of a material that is opaque to x-rays, such as uranium. This briefly traps the x-rays and channels them up the tube towards the secondary. The hohlraum walls don't actually reflect the x-rays but absorb them and quickly become so hot that they re-emit more x-rays.

The secondary is cylindrical in shape and multi-layered. The outermost layer is known as the pusher-tamper and is made of another x-ray absorber such as uranium-238 or lead. Inside that is a layer of fusion fuel. In the Sausage they used liquid deuterium and tritium, hence all the bulky cryogenic apparatus required. (Such a cumbersome device could never be deployed as a weapon so in later bombs lithium deuteride was used. This solid compound contains the deuterium needed for fusion and the lithium, when bombarded by neutrons produced during detonation, is transformed into the necessary tritium.) The innermost part of the secondary is known as the 'sparkplug' and is a hollow cylinder of fissile material such as plutonium-239 or uranium-235.

When the x-rays from the primary hit the pusher-tamper the outer layers are blasted off the surface at high speed causing the rest to recoil in towards the centre in what is called a radiation implosion. The imploding pusher-tamper compresses the fusion fuel to high pressure and that in turn squeezes the sparkplug. While it's a hollow cylinder, the sparkplug is not a critical mass but once the implosion squashes all the plutonium or uranium into the centre it reaches criticality and a second fission explosion starts. It's this second explosion, pushing outwards, that further compresses and heats the already highly compressed fusion fuel and gets the fusion burn started. The whole process takes a tiny fraction of a second and the resulting blast reduces the whole apparatus to atoms.

Although Teller had contributed to the breakthrough that made the H-bomb possible and was its greatest advocate, he was not chosen to lead the development effort that led to Ivy Mike, perhaps because of his reputation as a prickly personality. Feeling spurned by the Los Alamos lab, Teller moved in 1952 to the University of California, Berkeley, where with the help of Ernest O. Lawrence, director of the university's Radiation Laboratory, he founded a new weapons design lab in the nearby town of Livermore. The intention was to provide competition for Los Alamos and also to investigate some of the more way-out physics concepts that Teller was attracted to.

From the outset the Livermore lab focused on more innovative weapons design and, perhaps as a result, its first three nuclear tests were duds. But the lab went on to design many of the warheads that were manufactured in their thousands during the Cold War. In pursuit of these weapons, Teller and Lawrence pioneered the use of computers and computer simulation as a way of predicting how a bomb design will behave. The lab often owned the most powerful computer in the world and its designers became expert in devising simulations, which they called 'codes,' of nuclear explosions.

In the summer of 1955 a 24-year-old physicist named John Nuckolls left Columbia University in New York to join the thermonuclear explosives design division at Livermore where he was initiated into the secrets of the Teller-Ulam design and the use of weapon design codes. A couple of years later his boss asked him to look into an unusual scenario: If you could excavate a cavity inside a mountain 300m across – probably using a nuclear explosive – would it be economically viable to fill the cavity with steam and then set off a half-megaton H-bomb to drive the steam out and through a turbine to generate electricity? (Teller was a great enthusiast for devising peaceful uses for nuclear bombs.) Nuckolls estimated that the value of the electricity generated would cover the cost of creating the cavity, building the bombs and operating the facility, but he couldn't be sure how long the cavity would survive the repeated explosions. In any case, he couldn't see what advantage such a scheme would have over a fission power plant or even a magnetically confined fusion reactor.

But Nuckolls was intrigued by the idea and kept working on it. What if, he wondered, you reduced the size of the explosion so that it could take place in a smaller, man-made cavity? To do that you would need to find something other than an A-bomb to act as the detonator. An A-bomb requires, at the very minimum, a critical mass of fissile material to detonate; any less and it simply doesn't go off. 'Little Boy,' the bomb dropped on Hiroshima in 1945, contained 64 kilograms of uranium-235, not much more than its critical mass. So the smallest possible A-bomb detonator is still going to create a sizable blast. If some other way to set off a fusion explosive could be found, it would be possible to create much smaller explosions that could be contained in a controllable way. There is no critical mass for a fusion explosion; they can be as small as you want.

But how to produce the huge temperatures and pressures needed for fusion without the intense x-rays from a fission explosion? John Foster, head of another Livermore division dealing with fission weapon design, heard about Nuckolls' investigations and invited him to meetings of a special group he had set up to deal with that precise problem. One of the group, Ray Kidder, had already estimated the sort of conditions that would be required to ignite a small amount of deuterium-tritium fuel confined inside a metal capsule, or pusher – this is similar to the pusher in the Teller-Ulam design but spherical rather than cylindrical.

Nuckolls took away what he learned from this non-nuclear primary group and began to devise a scheme that could be used to explode tiny spherical capsules of D-T. He imagined many potential candidates for the energy source, or 'driver,' to set off the implosion, including a plasma jet, a hypervelocity pellet gun, and a pulsed beam of charged particles. Using weapons designers' codes he simulated a scenario where some driver caused the radiation implosion of a thin pusher capsule containing a tiny quantity – a millionth of a gram – of D-T fusion fuel.

The driver in his scheme would pump 6 million joules (6 MJ) of energy into the capsule in a pulse lasting just 10 billionths of a second (10 nanoseconds). The implosion squeezes the D-T fuel, raising its temperature to around 3 million °C. Nuckolls calculated that this would cause a burning fusion reaction in the fuel producing 50 MJ of energy, hence a gain of almost 10. Nuckolls had realised that compression was the key to getting the fusion to work. It would be possible to use the driver just to heat up the capsule, but imploding the fuel heats it up as a by-product. Nuckolls calculated that it was more energy efficient to heat it by compression, and by ending up with fuel that is hundreds of times as dense as lead you get many more collisions between ions that could result in fusion.

But Nuckolls knew that this first attempt wasn't good enough to be an energy source. For that he would need to achieve a gain of at least 100, because producing the energy pulse of the driver is likely to be an inefficient process. You may have to put 60 MJ into the driver to get a 6 MJ pulse, and converting the heat from the fusion reactions into electricity will involve more losses. A gain of 10 would certainly not be enough to come out with a profit. For a fusion power plant he would need a better design of fuel capsule, or target, able to produce more energy per explosion and cheap to manufacture – a commercial power plant would need lots of them. There were also exacting demands on the driver: it needed to produce high energy pulses of only a few nanoseconds' duration; it would have to focus its energy down to a tiny spot size – millimetres or less across – from a distance of some metres away so it wouldn't be damaged by the explosion; it would need to be efficient and low maintenance so it wasn't too expensive to run; and over the typical thirty-year life of a power plant it would need to ignite billions of explosions to produce economic quantities of power.

Nuckolls set about devising a more finessed scheme that would produce higher gain. His original target had a very thin metal shell acting as the pusher and this was coated on the outside with a layer of beryllium as an ablator, the material that absorbs the energy of the driver radiation and flies off, pushing the pusher inwards. One problem was due to instabilities: as the pusher moved inwards during the implosion, any slight irregularity in the pusher's thickness or in the force being applied by the ablator gets amplified and can end up with the pusher breaking up and allowing the pressurised fuel inside to escape. Another problem was the pusher itself which, being metal, weighed around a hundred times as much as the D-T gas it contained. Hence much of the energy of the driver was expended in accelerating the pusher rather than compressing the fuel.

So Nuckolls started doing simulations of targets made simply of a hollow sphere of frozen D-T fuel, dispensing with pushers and ablators entirely. Here the driver radiation falls directly on the surface of the D-T sphere and as some of it gets blown off it acts as its own ablator. With this sort of target it was not enough to have a strong pulse to get the pusher moving and then rely on its momentum to compress the fuel. With no pusher he had to tailor an extended pulse from the driver so that it keeps on pushing. The pulse would start off with low power and ramp up as the pressure within the imploding target increases. Nuckolls also manipulated the implosion so that the very centre of the compressed fuel got the hottest and the fusion burn would ignite there and then propagate outwards, consuming the rest of the D-T fuel.

From the amount of energy you could produce with such a fusion explosion, Nuckolls calculated that the targets would have to be very cheap, no more than a few cents each. A frozen D-T sphere might cost too much, so he simulated a scenario of using the equivalent of an eye-dropper to create droplets of liquid D-T. By further fine-tuning the length and shape of the driver pulse, Nuckolls was able to compress the droplet to a density of 1,000 grams/cm$^3$ – 100 times as dense as lead – and a central temperature of tens of millions of °C. It was a tour-de-force, but it remained only a simulation.

Nuckolls' fellow weapons designers didn't take his work very seriously. They referred to the numerous internal memos documenting his progress as 'Nuckolls' Nickel Novels.' In the strange world of the weapons labs, your designs aren't considered worth much unless they are turned into physical form, taken to Nevada or a South Pacific island and exploded. Nuckolls had no way to test his designs, so his colleagues thought of them as science fiction.

One of the key things that Nuckolls' schemes lacked was a driver, but the perfect thing was about to fall in his lap: the laser. Over the past few decades physicists had been studying the phenomenon of stimulated emission of electromagnetic radiation, such as light and microwaves. Stimulated emission occurs when one of the electrons around an atom is raised to a higher energy level and, rather than spontaneously jumping back down to a lower level, it hovers there briefly. Then if radiation with a particular wavelength should come along, its presence stimulates the electron to make the jump back down and emits its energy as more radiation. But this isn't just any old radiation; it's a perfectly minted copy of the wave that stimulated its creation – same direction, same wavelength and perfectly in step. Researchers realised that if you could somehow produce a large quantity of atoms with electrons in an elevated energy state, a small amount of radiation passing among them would quickly be joined by much more, all the same and in step – such radiation is said to be 'coherent' and is extremely useful.

In 1954 Charles Townes of Columbia University in New York with two graduate students succeeded in making a microwave amplifier using the stimulated emission principle. The excited material they used was ammonia gas and they named their device the 'maser,' an acronym for microwave amplification by stimulated emission of radiation. In Russia, Nikolay Basov and Alexander Prokhorov of the Lebedev Physical Institute in Moscow independently achieved the same feat. From then the race was on to do the same thing with visible light. At Columbia University, graduate student Gordon Gould jotted down some ideas for stimulated emission of light in November 1957, including the key idea of an open resonator – having the energised material sandwiched between two mirrors so that light bounces back and forth, stimulating many emissions before eventually escaping through an aperture as a narrow beam of coherent light. Gould had the great forethought of getting a notary to officially verify the date of his ideas. Townes meanwhile had teamed up with Arthur Schawlow of Bell Telephone Laboratories and a few months later in 1958 they filed a patent on a similar device and published a paper describing their ideas. Basov and Prokhorov were also closing in on an optical device and Prokhorov published a paper independently describing the open resonator concept the same year.

Gould presented his ideas at a conference in 1959 and coined the name 'laser,' using the same formula as for maser but amplifying light rather than microwaves. He filed a patent in April but it was rejected by the US Patent Office in favour of the rival patent from Bell Labs. This led to a bitter twenty-eight-year patent battle that was eventually won by Gould.

But in 1959, none of the competing groups was having much luck in building a working device. On 16th May, 1960, they were all beaten to the prize by Theodore Maiman, a physicist and electrical engineer whose doctorate concerned optical and microwave measurements of excited helium atoms. Inspired by Townes and

Schawlow's 1958 paper, he began to look for a suitable material for a laser medium. Working at Hughes Research Laboratories in Malibu, Maiman settled on synthetic ruby. He acquired a rod of ruby, put mirrors at either end and pumped it with light from flashlamps to create the necessary population of energised atoms. Because of the shortlived nature of the flashlamps, the ruby only produced short pulses, but the thin coherent beam of red light (wavelength 694 nanometres) had all the hallmarks of a laser.

Maiman announced his breakthrough in July and researchers at Livermore were instantly fascinated. The problems with existing sources of light was that the beams tended to diverge and the light contained a range of wavelengths, so when you tried to focus it with a lens onto a spot the different wavelengths focus at slightly different places and the spot is smeared out. A laser beam was as straight as an arrow and, as it contained only a single wavelength, all its light behaved in the same way and its energy could be focused onto a tiny spot.

Developments in laser research came thick and fast as scientists across the globe tried new laser materials and new pumping schemes to produce different wavelengths, demonstrated extremely short pulses (of the sort that would be necessary for fusion) and, most importantly, reached higher powers. By the spring of 1961 it was becoming clear that giant lasers might one day have enough power to drive radiation implosions. Laser light was ideal because the laser and all its focusing optics could be sited some metres away out of range of the explosion.

In September Nuckolls presented the idea of a 'thermonuclear engine' to John Foster, now the director of the Livermore lab. He described it as 'the fusion analog of the cyclic internal combustion engine,' with fusion capsules 'burned in a series of tiny contained explosions.' Nuckolls explained how lasers made all of this possible and rounded off his memo by suggesting 'possible applications for this engine are power production . . . or a thermonuclear rocket.' Whatever the merits of Nuckolls' plan it came at just the wrong time: global events soon meant that he had to put aside dreams of a fusion engine.

During the 1950s there was growing international concern about the rapidly accelerating nuclear arms race and the effect of radioactive fallout from testing in the atmosphere. In 1957 and early 1958 the United States and Soviet Union began to call for a moratorium on testing. A Conference of Experts was convened by the United Nations to investigate whether, if a treaty banned all nuclear tests, it would be possible to detect whether states were cheating and carrying out clandestine explosions. In August 1958 the conference reported that it would be possible to verify compliance with a network of 160 seismic monitoring stations spread around the globe. In October the United States, the Soviet Union and the United Kingdom – the three nuclear powers at the time – began negotiations in Geneva for a comprehensive test ban and agreed to stick to a one-year moratorium. The talks continued during 1959 but a major sticking point was over the extent to which nations could inspect each others' territory for evidence of secret tests. International tensions were increasing and at the end of the year, when the testing moratorium expired, it was not renewed, although none of the powers started to test again immediately.

France's first nuclear test, in the Sahara Desert in February 1960, further complicated the situation and the downing of an American U-2 spy plane over Russian territory in May meant there was no more progress in the negotiations that year. The new US administration of John F. Kennedy got the ball rolling again in March 1961 but when the US and UK put forward a draft treaty the Soviets again rejected the verification provisions. On 1st September, 1961, citing increased tensions and the French tests, Russia restarted nuclear testing. But it may have had another reason for abandoning the moratorium because two months later, on 30th October, Russia shocked the world by testing Tsar Bomba, the most powerful nuclear bomb ever detonated.

Soviet military strategists had a policy of developing the most powerful H-bombs that they could because they were lagging behind the US in the development of intercontinental ballistic missiles. Russia's first strike capability relied on long-range bombers which were inaccurate and vulnerable to getting shot down. So there was a logic to building big bombs to give the few bombers that got through as big an impact as possible, and if they missed the city centre they were aiming for by a few miles the city would still be devastated.

Even with that motivation, Tsar Bomba was not a practical weapon and was probably tested more as a show of strength to Russia's adversaries. The device was 8 metres long and 2 metres in diameter, larger than the 'Sausage' exploded on Elugelab, and weighed 27 tonnes. It was so large that a TU-95V bomber had to be specially modified to carry it. The design – a Teller-Ulam type – could in theory create a blast of 100 megatonnes (Mt) by surrounding the fusion fuel with a tamper made of uranium-238 so that neutrons from the fusion explosion trigger another fission explosion in the tamper. But in an effort to reduce the amount of radioactive fallout, the test replaced the uranium-238 with lead and this reduced the blast to 50 Mt – still ten times the total amount of conventional explosives used in World War II.

At around 11.30 on 30th October the TU-95V dropped the bomb above the island of Novaya Zemlya in the Arctic Sea. The bomb's fall was slowed by a parachute to give the bomber and its escort time to get a safe distance of forty-five kilometres away but still the shockwave caused the planes to drop in altitude by a kilometre. The fireball could be seen 1,000 kilometres away and the mushroom cloud rose to seven times the height of Everest. A village fifty-five kilometres away was completely destroyed and there was substantial damage in towns hundreds of kilometres away; windows were broken in Norway and Finland, more than 1,000 kilometres distant; and the seismic shock caused by the blast was still detectable after it had travelled around the world three times. Tsar Bomba was the most powerful device of any kind ever built. To produce such a blast with conventional explosives would require a cube of TNT 312 metres on each side, roughly the height of the Eiffel Tower.

At Livermore the explosion of Tsar Bomba was like a call of Action Stations! The US had to respond. After the brief pause of the previous few years everyone started preparing for new tests and Nuckolls had no time to devote to fusion engines. US testing resumed in April 1962 and there followed a furious six months of detonations by both sides. But the renaissance of testing was short-lived. Treaty negotiations resumed the following year and on 5th August the US, UK and the Soviet Union signed the Limited Test Ban Treaty which banned testing in the atmosphere, underwater or in space – so testing went underground. Nuckolls was kept busy designing nuclear weapons for the next few years, but not everyone at Livermore was directly involved in those efforts.

Ray Kidder, whom Nuckolls had worked with in the non-nuclear primary group, was keeping a close eye on the development of lasers, first by Maiman and others at Hughes, and then elsewhere. Hughes researchers devised a technique for producing ultra-short laser pulses and then others at the American Optical Company found that you could make lasers by mixing various so-called rare earth metals, such as neodymium, in glass. This was much cheaper than the ruby used by Maiman and opened the possibility of large rods and discs of laser glass.

Kidder carried out some rough calculations in late 1961 which suggested that a laser pulse with an energy of 100,000 joules (100 kJ) and lasting 10 nanoseconds might be enough to compress and ignite a small quantity of D-T fuel. The following April he visited Maiman at Hughes to discuss whether it was theoretically possible to scale up laser energy to hundreds of kilojoules. Reassured that there were no known obstacles, Kidder reported his results to Foster and soon he was heading the Livermore physics department's new Q Division, devoted to the interactions of electromagnetic radiation with matter.

Kidder and his colleagues began immersing themselves in the world of laser research and, in true Livermore style, Kidder devised a new simulation code optimised for laser-driven implosions. Using the new code in the summer of 1964, Kidder calculated that a laser pulse of at least 500 kJ lasting for 4 nanoseconds would be needed for ignition. After further refinement the estimate went up in 1966 to 3 million joules (MJ) in 5 nanoseconds. The estimated energy of the laser pulse required had gone up thirty times since Kidder's first calculation in 1962 and the pulse length had halved.

Most of the 1960s was spent learning about lasers and how their beams interact with matter at high energies. The key technologies developed during this period include methods of producing ultra-short pulses, with names like mode locking and Q-switching, and laser amplification. Researchers realised that after you had produced a beam using a laser you could shine the beam through some more lasing material, already pumped with energy but without mirrors at the end, so the beam gathers more light and energy as it passes through. In theory, a beam could be passed through dozens of these laser amplifiers to pump its energy up to a high level.

Kidder's team examined various laser materials to find which would be suitable for fusion and narrowed it down to two: iodine gas and neodymium glass. In the end they opted for neodymium glass (Nd:glass) because it was easier to work with. Neodymium ions made an ideal laser material. They could be pumped up with energy using relatively cheap xenon flashlamps and would stay in that state for 100 nanoseconds; plenty of time to form or amplify a laser pulse. Incorporating the neodymium into glass did cause some problems: often the refractive index of the glass – its ability to bend light – would vary which reduced the beam quality. So laser designers sought to use as little glass as possible in their designs. They would use sheets of Nd:glass 2.5 centimetres thick and shine light through it via the broad face. An amplifier was built up by arranging, say, sixteen disc-shaped sheets arranged in a row, each angled to the oncoming beam to prevent reflections. The sheets, since they were angled, could be pumped by flashlamps at the side.

By the 1970s they were able to take a seed pulse from an Nd:crystal laser, with an energy of 0.001 J, and pass it through a series of amplifiers with a combined glass thickness of 1 to 2 metres to produce a pulse with an energy measured in kilojoules – an increase in energy of more than a million times.

Also during the 1960s, it became obvious that the Livermore lab wasn't the only game in town. Its sister laboratory at Los Alamos was also investigating lasers for fusion, including ones using gases such as carbon dioxide and krypton fluoride as the laser material. Researchers there also looked into the possibility of using laser-ignited fusion capsules as a propulsion system for rockets. Researchers at the Naval Research Laboratory (NRL) in Washington, DC also began a research programme into lasers to see what military applications they might have. They and everyone else in the laser community were surprised by news in 1966 that researchers in France had produced a 500 J pulse with lasers and amplifiers of Nd:glass.

The best in the US at that time was less than 10 J. In 1967 NRL bought 60 J rod-shaped amplifiers from the French lab to study them and bought 500 J amplifiers in 1971. In 1967 Alan Kolb, head of plasma physics at NRL, heard a briefing in Washington about laser fusion by Kidder. Believing that NRL had superior skills with lasers than the two weapons labs, the naval researchers began their own programme of fusion research.

Then came news from Russia that Nikolay Basov, one of the co-inventors of the maser, and some colleagues had used an Nd:glass laser to heat a sample of lithium deuteride and had detected neutrons. Although it wasn't many neutrons, this was a sure sign that ions were undergoing fusion. Basov's report didn't suggest that the target was being compressed by the laser, only that it was being heated, so the team at Livermore were not too concerned that the Russians were ahead in what was rapidly becoming a race for laser fusion. What worried the Livermore researchers more was some documents circulating around the lab: they were patent applications from a commercial company claiming the invention of laser fusion through the radiation implosion of small fuel capsules. The US Patent Office had sent the applications to Livermore for review. If they were granted then this company, KMS Industries, would control this new technology, not the government labs that felt it was their private reserve.

The author of the patents was Keith Brueckner, a professor of physics at the University of California, San Diego. Brueckner had a finger in very many pies: he had done consulting work for the Los Alamos weapons lab, the Department of Defense and the Atomic Energy Commission. He had studied lasers and in the late 1960s was an adviser to the AEC on controlled fusion, having studied magnetically confined fusion at Los Alamos a decade earlier. In 1968 the AEC asked Brueckner to attend the International Atomic Energy Agency fusion conference in Novosibirsk as their representative. This was the meeting at which Lev Artsimovich sparked the race for tokamaks by revealing his results from the T-3. But Brueckner wasn't there to find out about tokamaks; he was sent to find out what other countries were doing in laser fusion. He listened to reports from Russian, French and Italian researchers and mixed with them socially afterwards. Returning home, Brueckner wrote up a report for the AEC on what he had learned in Novosibirsk. He had become intrigued by the topic and talked informally with various AEC officials to see if the commission would fund him to carry out a more thorough investigation. He received no encouragement, presumably because, although Brueckner didn't know it, the AEC already supported all the work in this area at Livermore and Los Alamos.

Undaunted, Brueckner applied to the Department of Defense and was given a small (classified) research contract to investigate laser fusion. He started to study the idea of focusing the beams from a number of powerful lasers onto a small sphere of mixed deuterium and tritium and heating it enough to start fusion reactions. Normally, the heated plasma would rapidly expand but, because of its inertia, it would take a moment to do so. Like similar schemes, he was relying on that moment of inertia to be long enough for the plasma to reach fusion temperature. So the heating has to be very fast, requiring an ultra-short, powerful laser pulse. Because of this reliance on inertia, laser fusion is often called inertial confinement fusion.

Not knowing about Nuckolls' detailed simulations, Brueckner had only found rough calculations of the sort of pulse needed to achieve fusion. So, helped by some other theorists at UC San Diego, he developed simulation codes, taking account of the energy deposited by the laser beams, the conduction of heat in the plasma and its expansion, shock waves and the effect of any fusion reactions that occurred. In 1969 the code was finished and they started running some simulations. To their amazement the codes predicted that they would produce huge numbers of neutrons, hundreds or more times as many as they had expected based on estimates from other people's work.

They analysed in detail the output of the simulation and realised what had happened: rather than just heating the plasma the laser pulses where compressing the capsule and heating it too. Almost by accident they had stumbled on the idea of using a radiation implosion to trigger fusion. Brueckner's calculations showed that it was possible to get much more energy out of a capsule using implosions and that they could achieve ignition with a laser pulse of only a few kilojoules, not much more than state-of-the-art lasers could produce at the time.

It was around this time that Keeve 'Kip' M. Siegel, head of KMS Industries, became involved. In addition to all his government consulting work, Brueckner had, for the previous few years, been consulting for KMS Industries and when his meagre DoD funding for laser fusion proved inadequate Siegel's company had provided some additional cash. Brueckner told Siegel about his simulation results. The industrialist was hugely enthusiastic and determined that this was something that KMS Industries had to be involved in.

Kip Siegel was the quintessential American scientist-entrepreneur. Trained as a physicist, he became a professor of electrical engineering at the University of Michigan at Ann Arbor. Much of his work there was concerned with identifying aircraft and missiles using radar. He patented some of his inventions in the fields of radar and electro-optics and in 1960 set up a company, Conductron, to market them. Over six years he turned the company into one of the area's biggest industries before finally selling it to McDonnell-Douglas for $4 million. Soon after he formed KMS Industries and built it up into a nationwide conglomerate by buying up other small high-tech companies.

In spring 1969 Brueckner and Siegel went to visit Paul

McDaniel, head of research at the Atomic Energy Commission in Washington. Because some of the work had been supported by KMS Industries, they asked McDaniel if he would treat the information they were about to give him as proprietary secrets. He said that if that was the case they should go away again and apply for patents before talking to anyone at the AEC. So they returned to California. Brueckner wrote up his results and applied for his first patent in the summer.

Later in the summer Brueckner was in West Palm Beach in Florida for a meeting when he had a visit from the AEC's head of security who told him that the work he was doing on laser fusion was related to classified weapons design work and that he was forbidden to do any more experiments or simulations, talk to anyone about his work, or even to do any calculations on paper. Brueckner soon learned that the Patent Office had, as a matter of routine, sent his patent applications out for comment to experts in the field, in this case to Livermore, Los Alamos and the AEC. Their contents had caused uproar, hence the crackdown by the AEC on any further work.

The AEC's reaction only served to convince Siegel even more that they were onto something important. Siegel got his lawyers to work on the AEC and eventually a compromise was struck whereby Brueckner alone could continue to work on laser fusion. During that autumn and into spring 1970, Brueckner continued his theoretical work and wrote up more patent applications, eventually producing a total of twenty-four. He was so annoyed by the AEC's draconian restrictions that he made as many and as wide patent claims as he could. Meanwhile, Siegel's Washington lawyers continued to lobby the AEC until finally in the spring it allowed KMS Industries to continue work on laser fusion. However, the research remained classified and under government control, KMS Industries would receive no funding, get no access to AEC research and could not hire former AEC employees.

Siegel thought the work could be done with KMS Industries' own resources and help from other industrial partners. While Brueckner and his colleagues at San Diego continued doing theoretical work, Siegel made plans and in spring 1971 announced that they would build a laboratory in Ann Arbor, raising money by selling off some divisions of KMS Industries. To entice Brueckner, Siegel offered him a significant fraction of any profits that came from the work, so Brueckner took leave from UC San Diego and moved to Michigan.

The race for laser fusion was rapidly heating up. Evidence was emerging that Basov and his colleagues at the Lebedev Physical Institute had also discovered the importance of imploding the fusion target and a university-based effort in the United States was also joining the fray. This was the brainchild of Moshe Lubin, an Israeli researcher who came to the US in the early 1960s to study aeronautical engineering at Cornell University. In 1964 he joined the department of mechanical and aerospace sciences at the University of Rochester in New York State. Lubin was fascinated by the rapidly developing science of lasers, particularly the possibility of focusing beams down to produce high energy densities and the effect this might have on matter. His department at Rochester was fertile ground because there were already others there interested in plasma physics and the university had an Institute of Optics. Rochester is also the home of the Eastman-Kodak camera company which provided hardware for their laser development and studies of laser-matter interactions.

By 1970 Lubin and his colleagues believed that lasers could heat a plasma sufficiently to ignite fusion reactions and in the autumn the university set up the Laboratory for Laser Energetics (LLE) with Lubin as its director. Lubin set out his manifesto in a 1971 article in *Scientific American*, describing heating fusion targets with Nd:glass lasers. He predicted energy break-even with a laser pulse of less than 1 MJ and even described a reactor vessel with thick liquid-lithium walls to absorb the neutrons. That year LLE began work on Delta, a 1-kJ laser with four separate beamlines designed to carry out fusion experiments.

The researchers at Livermore knew they had to respond to all this competition but were hindered by the slow-moving bureaucracy of the lab and the AEC. Nuckolls now had more time away from weapons design work and was again getting involved in fusion target design, teaming up with a young protégé of Teller's called Lowell Wood. They lobbied hard in the lab and at the AEC for aggressive laser-fusion development, working towards a 10-kJ laser. Reports from Russia said researchers there had already achieved implosions using lasers with multiple beams while the teams at Rochester and the newly established KMS Fusion were already building similar lasers. While the researchers at Livermore had done huge amounts of simulation and investigated the properties of various laser systems, they had yet to fire a pulse at a fusion target.

Nuckolls and Wood favoured building an Nd:glass laser which produces infrared light with a wavelength of one millionth of a metre (1 micron). Some pushed for the carbon dioxide gas lasers with a 10-micron wavelength being developed at Los Alamos. Although these were much more energy efficient than Nd:glass, it was believed that the shorter wavelength was absorbed more readily by the target and was less likely to be affected by plasma blasted off the target during the drive pulse. Others, including Kidder and Teller, thought that not enough was yet known about high-power lasers and so the research should progress more slowly, but the lab managers saw that lasers had wider uses in weapons physics and other military and industrial applications, so they backed the programme. The support of the AEC was won in early 1973 and laser building began.

Meanwhile, the AEC had also allowed some declassification of the laser-fusion work being done in the weapons labs. Nuckolls argued strongly for the declassification of his scheme for using a bare drop of liquid D-T directly driven by lasers. Because no hohlraum is used, he argued, it wouldn't reveal any secrets about weapon design, nor would the manufacturing details of precision fusion targets be out there because all that was required was an eye-dropper. The AEC acquiesced and this paved the way for the May 1972 International Quantum Electronics Conference in Montreal at which Nuckolls, Wood, Teller and others from Livermore presented much previously classified material. Basov led a delegation of Soviet researchers and other groups also discussed their work. The conference opened up laser fusion to a much wider scientific audience and played a similar role to that of the 1958 Geneva conference for magnetic fusion researchers. Nuckolls followed that in September with a now-famous paper in the journal *Nature* which gave more details of the directly-driven bare drop design and other declassified information. For many scientists around the world, this was their first detailed introduction to laser fusion.

Livermore recruited new people for its rapidly expanding laser programme, including John Emmett from the Naval Research Laboratory who was an expert in Nd:glass lasers and would lead the programme. First, in 1973, they built the 100-J Cyclops just to gain experience in highly amplified lasers. They had to overcome many problems, including flashlamps that exploded and dust on the surfaces of glass discs of the amplifiers – the high-powered beam would heat up the dust, which would in turn damage the surface of the disc. Next, in 1974, came the 20-J Janus which, although a lower energy than Cyclops, had a much shorter pulse length (0.1 nanoseconds) to match the timing of target implosions.

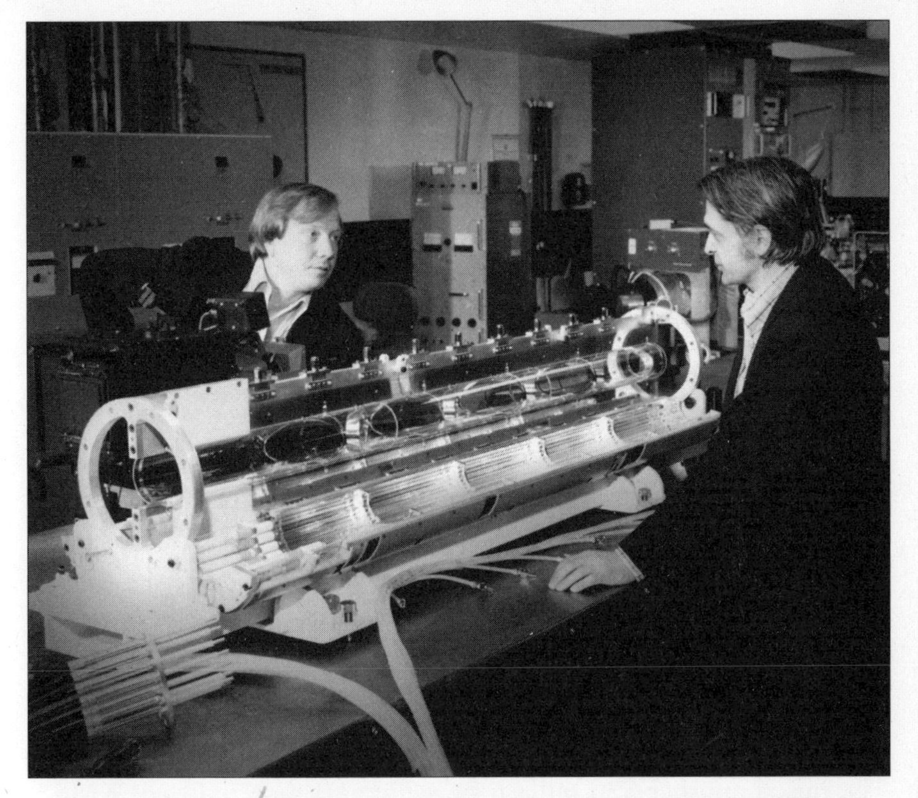

**John Emmett (left) and John Nuckolls with a laser amplifier.**
(Courtesy of Lawrence Livermore National Laboratory)

The first beamline was soon joined by a second to take the energy to 40 J. Funding for the project began to increase rapidly around this time, in part because of the Yom Kippur War and the Middle East oil embargo that followed. Money was poured into energy projects following that first oil crisis and, although laser fusion was never given the label of an energy project, more money was channelled into it via defence spending.

Despite the money and resources that the government labs had at their disposal, they were not the first to observe fusion neutrons by imploding a target with a laser. That prize went to KMS Fusion. Brueckner had spent a frantic three years setting up his laboratory from scratch. He had hired eighty scientists and technical staff, contracted General Electric to make the laser to his specifications, modified the buildings Siegel provided in Ann Arbor and installed the laser. His team devised a clever way of making fusion targets. They made tiny micro-balloons with walls of extremely thin glass. To fill them with deuterium-tritium fuel they surrounded the micro-balloons with the gas mixture at high pressure of more than fifty atmospheres and heated them to 500 °C. At high temperature the glass becomes permeable and the D-T gases can diffuse inside. They then cooled the balloons and the fuel became trapped. If desired, the capsules could be chilled to extremely low temperature to condense the gases and create a thin layer of D-T ice on the inside of the glass. Meanwhile, Siegel had been working tirelessly to raise money for the research. He was wildly enthusiastic about fusion. He talked about fusion reactors small enough to fit into a garage that could provide power for a small community. To show off KMS Fusion's technical skills his lobbyist in Washington would hand out medicine vials each containing 10 million micro-balloons. He secured $8 million from the British company Burmah Oil and he raised another $20 million by selling off more parts of KMS Industries but it proved hard to persuade other investors to pitch in until Siegel's hyperbole was transformed into real results.

Those results came in 1974 when they began imploding their glass targets with the new GE-built laser. As they only had one laser, Brueckner's team reflected its light off two specially designed elliptical mirrors to produce an even distribution of light all over the target sphere. The laser pulses – delivering around 100 joules in 0.3 nanoseconds – compressed the D-T fuel to ten times its normal density as a solid and, to the delight of the team, produced lots of neutrons, often as many as 7 million from a single shot. Siegel, understandably, trumpeted the breakthrough far and wide. The press and scientists had been starting to grow weary of his frequent announcements of progress and scepticism had crept in. The neutron results served to silence those critics.

Although KMS Fusion's results were an unmistakable sign of fusion reactions taking place it was still a long way from a useful amount of energy. Each target would have to produce ten million times as many neutrons just to break-even with the amount of energy in the laser pulse. But it was a major proof-of-principle for laser fusion which thus far in the United States had existed mostly in computer simulations. That year the AEC slightly loosened the bounds of classification again so that researchers could talk publicly about hollow targets as well as Nuckolls' bare drop model. As a result, Brueckner went to the American Physical Society meeting in Albuquerque, New Mexico, in October; was allowed to describe his experiments; and was duly lauded by his peers. To the frustration of researchers from Livermore, this commercial upstart seemed to be leading in what some were calling the 'neutron derby.' To further rub salt into their wounds, in March 1975 the Energy Research and Development Administration (ERDA), the new agency that took over the research parts of the AEC when it was wound down earlier in the year, awarded KMS Fusion a contract worth $350,000 to carry out shots on behalf of the Livermore researchers to validate their computer simulations.

Despite these embarrassments, the Livermore team was starting to make some progress. Its Janus laser, which had two beams so that light could be directed at the target from two directions simultaneously, was beginning to produce results. Similarly, Rochester's LLE had finished and was now operating its new laser. Called Delta, it had four beams and could produce a pulse with 1 kilojoule of energy. Now at both labs and at KMS Fusion, their years of simulating and predicting were butting up against the hard reality of experiments – and they were getting some unpleasant surprises. The first of these was plasma instabilities: when the laser beams hit the material of a micro-balloon target, the material that is kicked off the surface forms a cloud of plasma and this interacts with the beams in unpredictable ways. The plasma can scatter the incoming beams, preventing some of the laser energy from doing its job of imploding the target. The laser beams also heat up the electrons in the plasma and the superheated electrons can penetrate into the micro-balloon, heating up the fuel before the implosion gets started – high temperature fuel is much harder to compress than cool fuel.

Teller had predicted there would be such problems. In the late 1960s he was listening to a presentation at Livermore by one of the laser fusion researchers who was describing how the beams would produce a plasma around the target. As the explanation proceeded, the scowl on Teller's face deepened and deepened. Teller had been closely involved in the early days of magnetic confinement fusion and knew how devilish plasma could be. 'Wait a minute. Wait a minute! Are you telling me that laser fusion involves REAL plasma physics?' Teller asked. 'Yes, sir, it does,' replied the speaker. 'Well,' Teller said, with an air of disappointment, 'it will never work.'

When they encountered that real plasma physics, the teams of researchers concluded that they just needed more powerful beams to overcome the energy-dissipating effects of the plasma. More powerful beams meant new larger lasers. The lasers that they were working with – Delta, Janus and so on – were already big machines. With their master lasers, multiple amplifiers, optics to shape and condition the beams, and target chambers, each filled a large room. The next generation would require purpose-built experimental halls and cost tens of millions of dollars. At Rochester, Lubin had a vision of following Delta with a huge 24-beam Nd:glass laser called Omega but funding it would be a problem. The LLE to date had been funded by the university, industry sponsors and New York State, but the sort of money they needed for Omega would require federal funding.

Unlike its predecessor the AEC, the ERDA was outward looking and wanted its research to take place not just in the national laboratories but in industry and universities. Rochester saw this as an opportunity to get its foot in the door and get a slice of government funding. And, for the first time, Congress' Joint Committee on Atomic Energy was holding a session entirely devoted to inertial confinement fusion and how to divide up ERDA's budget for it in fiscal year 1976. Rochester University pulled strings in Washington to get Lubin a slot in front of the committee so that he could make a pitch for funding for Omega. But there would be a lot of competition for funds. Livermore had already started to work on a powerful new Nd:glass laser, the 20-beam Shiva. Other national labs were also due to make requests, as was KMS Fusion.

At 2 p.m. on Thursday, 13th March, 1975, Senator Joseph Montoya called the session to order. He began by reminding the committee that the President's 1976 budget request called for $212 million for fusion research, $144 million for magnetic confinement fusion and $68 million for inertial confinement, and pointed out that the US government had since the beginning of the programme in the early 1950s spent a total of around $1 billion on fusion research. He continued:

> We know that Fermi achieved criticality in the first reactor only 3 years after the discovery of the fission process. We recognize the engineering genius that put Americans on the Moon and brought them home, but we also recognize that Mother Nature has been very reluctant to give up her peaceful thermonuclear secrets. This is why the Joint Committee and the Congress have been willing to support two different approaches to fusion.
>
> Continued support, it seems to me, must be predicated on careful, step-by-step research programmes. Experience has shown to date that there is no quick solution to controlled nuclear fusion.

With those cautionary words, the session began by hearing from Major General Ernest Graves, head of ERDA's division of military applications which funded inertial confinement fusion. Graves noted that six labs had so far achieved implosion of a fuel pellet and some of those detected neutrons. The six were Livermore, Los Alamos, Rochester's LLE and KMS Fusion, plus labs in France and Russia. He then listed a series of milestones which, with healthy levels of funding, the programme should be able to achieve. These included 'significant thermonuclear burn' – in other words an implosion in which several percent of the D-T fuel fuses – by 1977-78; 'scientific break-even' – energy output equals energy of laser pulse – between 1979 and 1981; and 'net energy gain' between 1981 and 1983. If these milestones were successfully met, he predicted a 'test system' would be operating by the mid 1980s and a 'demonstration commercial plant' by the mid 1990s. But the key to getting there, he said, was laser energy. 'The pace of the overall programme depends on the rate of laser development and construction. This, in turn, depends upon the ingenuity of man and the level of funding,' Graves told the committee.

There followed talks by the directors of Los Alamos, describing their work on carbon dioxide gas lasers, and Livermore, discussing Shiva. The director of Sandia Laboratories spoke of their work on using beams of electrons instead of lasers to spark fusion. His laboratory was looking to build a new accelerator to test the feasibility of e-beam fusion. Then it was the turn of Moshe Lubin. He highlighted LLE's achievements so far: implosion experiments achieving densities of thirty times that of a solid, neutrons detected, and the world's only laser that can produce pulses with energies greater than a kilojoule and shorter than a nanosecond. All achieved without relying on federal funding.

To move forward towards break-even, LLE wanted to build a new laser capable of 10 kilojoule pulses. Lubin estimated that this would cost $40 million over six years: $24 million from industrial sponsors for the operation of the new facility; $6 million from the State of New York for the laboratory building; and, he proposed, $10 million from ERDA for the laser itself. In contrast to the presentations from the national laboratories and their still partly classified research, Lubin painted a picture of an open user facility which researchers from all over the country could come to and use. He likened Rochester's position in laser fusion to that of Princeton University in magnetic confinement fusion and said that in this early, fundamental phase of laser fusion research it was appropriate that 'the probing environment and backing of significant research capabilities of a leading university should be selected as the logical place for a major open research facility.' Lubin was taking a big risk: this was a field of research dominated by the large and influential weapons labs and supported by ERDA's division of military applications and he was saying that one of its major research facilities should be placed in the open, academic atmosphere of a university.

The final speaker was Kip Siegel. Siegel was in a tight corner. He had so far spent $20 million on KMS Fusion, cannibalising some forty-two divisions of the parent company KMS Industries to pay for the Ann Arbor laboratory which was by then one of the world leaders in laser fusion. But Brueckner and the KMS Fusion team had discovered, like their competitors at Livermore and Rochester, that compressing a fusion target was much more complicated than they had expected and overcoming the instabilities and laser-plasma interactions would require bigger and much more expensive lasers. Brueckner could not see how Siegel could afford it and so in autumn 1974 he had left KMS Fusion and returned to UC San Diego. Siegel was now without his key scientist and without the money to keep up with the government-funded fusion labs. He really needed to pull something out of the hat and that's exactly what he did, although that was not what most people remember about his testimony that afternoon.

A natural self-publicist, he surprised all assembled by stating that he was not there to talk about a laser fusion research programme. 'What I am talking about today is a possibility of having a pilot plant in existence in 1979 and 1980 producing hydrogen or methane, utilizing a laser fusion reaction. We visualize methane going into the pipeline in 1985, starting to make up for the shortfall in natural gas that will exist at that time,' he said. Siegel told the committee that KMS Fusion had, like all its competitors, assumed that the thing to do with all the high energy neutrons coming out of a fusion reactor was to use them to generate electricity. He said they got past this 'mental block' when the head of the Texas Gas Transmission Corporation came to him and asked: 'Can't you do something in fusion to produce gas?' Texas Gas was facing a drop-off in natural gas reserves and was looking to find an alternative source to take advantage of all the pipelines that it already had installed. KMS Fusion did some experiments and, according to Siegel, found a way to produce hydrogen using neutrons. That hydrogen could be used to make methane at less than half the cost of other synthetic gas methods such as coal gasification.

Siegel gave no details of his technique but it is possible that if aiming to use fusion neutrons in a chemical process rather than to generate electricity, you may not need such a high level of gain to make the process viable. This may explain Siegel's extremely aggressive timetable for building a demonstration plant in just four or five years. He asked ERDA for funding or a loan of $60 million over three years for the plant, to be added to $15 million from KMS Industries and $40 million from Texas Gas. As he neared the end of his testimony Siegel suddenly stopped in mid-sentence. The hushed audience waited for him to continue but he uttered the word 'stroke' and collapsed. The session was halted as an ambulance took Siegel to the George Washington University Hospital. He died at 5 a.m. the following morning and his dream of generating fusion energy in the private sector, and of making gas using neutrons, died with him. KMS Fusion continued to work in the field but was never again a big player in laser fusion. Rochester, in contrast, won the funding it was seeking and a year later, in April 1976, a cornerstone-laying ceremony was held to mark the start of construction of the Omega laser building.

There was another event that spring which would, in time, have a profound effect on laser fusion research: another international treaty limiting tests of nuclear weapons. Nations were already restricted to only testing weapons underground, but the Threshold Test Ban Treaty, which came into effect in March, abolished any tests of greater than 150 kilotonnes. This made it virtually impossible to design any new weapons in the megatonnes range or even to check that existing ones were working.

One of the rationales for carrying out nuclear explosions was to test the effects of the blasts on satellites, warheads and other military hardware. The Pentagon expected that politicians would eventually agree on a total ban on nuclear tests, so at that time it was pouring millions of dollars into large facilities – including particle accelerators, nuclear reactors and electromagnetic pulse generators – that could simulate the radiation from nuclear explosions. The problem with these facilities is that each could only produce one type of radiation, such as x-rays, gamma-rays or neutrons. What they needed was something that could produce all of them at once, just like a real explosion – something like a nuclear weapon in miniature.

A decade and a half earlier, weapons designers at Livermore had mocked Nuckolls' ideas for laser fusion. Now they were beginning to get interested. Laser fusion's tiny blasts really are like mini H-bombs and produce all the radiation that a full-sized bomb would. To test the radiation toughness of military hardware under reasonably realistic conditions all they had to do was open a port inside a fusion facility's reaction chamber and attach the piece of hardware to the wall. The blasts didn't even need to achieve high gain; all that was required was lots of neutrons and high-energy radiation. In addition, the tiny blasts could be used to study the 'weapons physics' at work in a bomb explosion and to provide real data to validate the computer simulations that designers used. In its last months of existence towards the end of 1974, the Atomic Energy Commission had initiated its first study of the prospects of laser fusion. Its conclusions were delivered to ERDA in March 1975 and it recommended 'aggressive development' of laser fusion technology and that the national research programme should be broadened out from the national laboratories into industry and universities. But the study also concluded that there were many difficulties to be overcome before laser fusion could be a viable energy source. Much more likely, the panel said, was that in the short term at least laser fusion would prove very useful in nuclear weapons simulation. From that time onwards there would always be a question mark over laser fusion research: is it an energy programme or a weapons programme?

Livermore's Shiva laser was finished in 1977 and was slowly ramped up to its full energy of 10 kilojoules in its first year. Although the laser was a success, the physics of pumping laser power into a tiny fusion target did not turn out to be as simple as the lab's computer simulations had suggested. The following year, Rochester started using the first six Omega beamlines in a proof-of-design setup they called Zeta. Researchers there soon encountered the same problems. Apart from the laser-plasma interactions encountered by earlier lower-energy experiments, another problem cropped up too: once the implosion got started the fuel inside the capsule didn't want to stay in there and had a habit of breaking through the ablator and squirting out in unpredictable directions.

This phenomenon, called Rayleigh-Taylor (RT) instability, is best understood by imaging a dish containing a layer of oil with a layer of water on top of it. This is an unstable situation because water is the denser liquid and, because of gravity, would naturally sit at the bottom but the oil is in its way. The two layers could remain as they are but if there is any ripple or momentary irregularity in the thickness of the water, a 'finger' of oil will push upwards through it and there will be a similar downwards thrust of water and the two layers will rapidly change places. In an imploding capsule, the dense ablator is the water, the D-T fuel is the oil and the applied force is the implosion driver. What researchers wanted was for the unstable denser-on-top arrangement to hold until the fuel is compressed, but that requires no irregularities in the ablator layer or in the applied pressure of the driver which would give the fuel an opportunity to burst up and out.

Making the ablator layer absolutely smooth and symmetrical is a matter of better manufacturing. Making the driver absolutely even is much trickier. As laser facilities had got bigger they had sprouted many more beams. Shiva boasted twenty beams and Rochester's Omega would have twenty-four. This meant that they could shine beams at the target from many directions to get an even overall coverage. But that was not enough to prevent RT instabilities. If you imagine looking at a laser beam end on, coming straight towards you, because of various optical effects the beam does not have an even intensity across its face and when the beam is focused down to a tiny spot on the target's surface that unevenness allows RT instabilities to develop.

Livermore's answer to the RT instability problem was to use hohlraums, the radiation cases like tiny tin cans that are modelled on the much larger hohlraums used in H-bombs, because then you don't have to fire the laser beams onto the target at all. In fusion experiments a hohlraum is about the size of the eraser at the end of a pencil and is made of a heavy metal such as gold. The

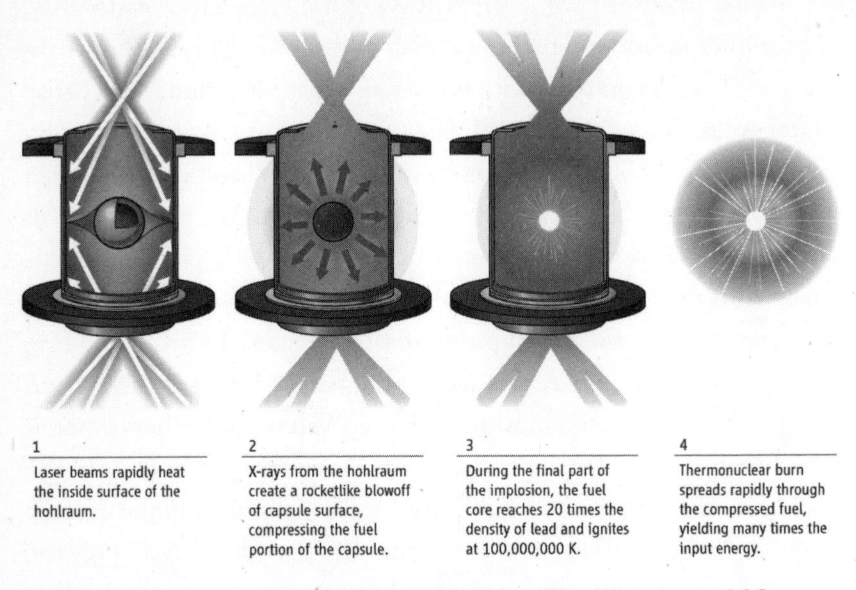


**Indirect drive at the National Ignition Facility: how 192 laser beams cause a fusion reaction.**
(Courtesy of M. Twombly/*Science*. Reprinted with permission from AAAS.)

hohlraum has holes at either end and the spherical fusion capsule is held at its centre. In a laser shot the beams are fired in through the hole at each end at an angle so that they don't touch the capsule but instead hit the hohlraum's inside wall. The laser pulse heats the gold of the hohlraum to such a high temperature that it emits a pulse of x-rays and it is these that cause the capsule's ablator to blow off and implode the fuel. The idea is that the conversion from laser light to x-rays smoothes over any unevenness in the laser beams. But this technique, known as indirect drive, does have drawbacks. Firstly the hohlraum, being cylindrical in shape, is far from symmetrical, so you have to be very clever with your placement of laser beams to get uniform x-ray illumination of the target. It's also inefficient: much of the laser energy is lost when the light is converted into x-rays. And indirect drive does not get around the problem of hot electrons and plasma interfering with the incoming laser beams; now it was plasma blowing off the hohlraum that was getting in the way. Despite these problems, the Livermore researchers had worked with hohlraums for years, understood them and believed that they could make them work.

The Rochester researchers, by contrast, mostly had backgrounds in optics and lasers rather than weapons design, so their inclination was to try to fix the unevenness of the laser beams. Shining laser beams straight onto the target – known as direct drive – was simpler and more efficient since much more of your beam energy goes into the implosion. As a result the laser fusion community in the United States split into two camps, with direct drive proponents led by Rochester and the Naval Research Lab while Livermore became the champion of indirect drive.

To combat the problems of laser-plasma interactions and hot electrons, researchers were beginning to realise that what they needed were new lasers with shorter wavelengths. Experiments in France had shown that laser beams with short wavelengths are absorbed better by targets and caused less pre-heating of the fuel by electrons. Rochester began a concerted effort to find other types of laser, but none could produce the right combination of high energy, short pulses and short wavelength.

However, during 1979 and 1980, Rochester researchers came up with the next best thing: a way to convert the infrared light from Nd:glass lasers to a shorter wavelength. They used a crystal called potassium dihydrogen phosphate, or KDP, which has a very useful property: if a beam of light is shone into it, the crystal structure of KDP interacts with photons of light in such a way that two photons can be merged to create a new photon with twice the energy and hence half the wavelength. This phenomenon only takes place in certain crystals and with light of high intensity, so it wasn't observed until 1961, just after the discovery of the laser. The laser fusion researchers at Rochester could thus convert the 1.054-micrometre light from their Nd:glass lasers to 0.532 micrometres, green visible light. The problem was that it only converted 43% of the beam's energy to the shorter wavelength, but the theorists at Rochester worked out a more elaborate scheme. After the first conversion they passed the new green beam and the remaining unconverted infrared beam into a second KDP crystal. Here the different photons combine to create new ones with three times the energy of the original ones and hence one-third of the wavelength, 0.351 micrometres, which is in the ultraviolet. The researchers found that if they carefully set the crystal orientations and the polarisation of the light in a particular way they could convert 80% of the original beam energy into ultraviolet light.

This opened up exciting new possibilities and LLE experimentalists started investigating how ultraviolet light would work as a fusion driver. But soon their thoughts were elsewhere: Moshe Lubin, LLE's leader and driving force, resigned to take up a job in industry. Standard Oil of Ohio, which was one of LLE's main corporate sponsors, offered him the post of vice president for research and it was too tempting to refuse. Jay Eastman, the chief engineer who had spearheaded the building of Omega, became LLE's new director but he also stepped down in late 1982 to set up a company making barcode scanners. Just when the lab had a new technique that showed such promise, it found itself in administrative turmoil.

From their experiments with UV light it became clear that laser fusion research at Rochester was not going anywhere unless they added a KDP-crystal conversion system to all twenty-four beams of the new Omega laser. It was a difficult task going back to the Department of Energy – which had taken over responsibility for fusion research in 1977 from the short-lived ERDA – to admit that their brand-new Omega laser wasn't working as predicted and that they needed to install expensive wavelength converters to it. Following much negotiation, the DoE finally agreed to a phased conversion over three years but it insisted that some economies had to be made at the lab. As conversion of the twenty-four beams began, lab managers had to cut more than 20% of the staff – an unpleasant process in the small, tightly knit LLE team; and some believed that LLE would never recover.

Laser fusion as a whole had hit a rough patch. Expanding at breakneck speed during the 1970s, spurred on by the oil crisis, the US government spent $1 billion on the field during that decade. At the beginning of the 1980s things looked very different: not only was oil cheap but the problems that the labs were having getting targets to implode successfully did not make it look like energy break-even was going to arrive any time soon. General Graves' prediction to Congress in 1975 that break-even would be achieved around 1980 now looked hopelessly optimistic. In 1981, the government's budget request for laser fusion research went down for the first time.

At the beginning of 1983, Robert McCrory was appointed as the new director of LLE. A veteran of Los Alamos, he had joined LLE in 1976 just as the construction of Omega was starting. Now he brought a measure of stability to the lab and it gradually got back on its feet. A few months after his appointment, the conversion of the first six beamlines to UV operation was finished and all twenty-four beamlines were converted by the autumn of 1985. The researchers now had to learn how implosion worked all over again with the new shorter wavelength of light. While the UV beams helped with reducing laser-plasma interactions and getting more energy onto the target, the team still had to grapple with RT instabilities – how to ensure that the implosion proceeded smoothly and symmetrically. Shining the beam directly onto the target had to be a better system than the complexity of indirect drive; they just had to work on their lasers to make them more uniform.

The Rochester team came to realise part of the problem was that laser beams were *too perfect*. Because the waves are all perfectly in step and all one wavelength, any imperfection in the optics is simply propagated along with the beam to the end target. What they needed to do was rough it up a bit – a bit of fuzziness in the beam's properties might help smooth over imperfections. One way they did this was to add an optical device at the end of the beamline called a distributed phase plate. This chops the beam into 1,500 beamlets and, while doing so, applies small random time delays to each one. The beamlets are then focused onto the target but, instead of making 1,500 tiny spots, they are all superimposed onto the same spot. Because all of these beamlets are now slightly out of phase with each other, there is an averaged-out illumination over the whole of the spot, which covers up imperfections. Another technique they developed, called smoothing by spectral dispersion, did a similar job with the beam's wavelength – roughing it up to cover over non-uniformities.

With the UV light and the new beam smoothing techniques, the LLE researchers were by 1988 able to compress targets to between 100 and 200 as dense as liquid D-T, a goal set for them by DoE in 1986. They were, in their mind, well on the way to demonstrating direct-drive laser fusion as a viable source of energy.

At Livermore in the late 1970s, Shiva was struggling. Because of the unforeseen complications of laser-plasma interactions, hot electrons and RT instabilities it was not performing as the researchers' computer codes had predicted. What the Livermore researchers did have on their side was very good diagnostic equipment to figure out what was going on during the implosions. They measured, they tweaked their models, and they made modifications. They adjusted the focusing of the beams into the hohlraum, increased the hohlraum size and changed the shape of the pulse.

In the end they were able to achieve the nominal goal of compressing targets to 100 times the density of liquid D-T, but the amount of neutrons produced was much less than predicted and they remained very far from energy break-even.

In time-honoured fashion, their solution to the problem was to build another, more powerful laser. Again using Nd:glass technology producing infrared light, the proposed Nova laser would generate twenty beams with a total energy of 200 kilojoules (up from Shiva's ten kilojoules) and cost some $200 million. John Nuckolls was convinced that this one would achieve ignition.

In 1979 the DoE assembled another panel to make an assessment of its inertial confinement fusion programme. Headed by John Foster, a former Livermore director, it had to consider, among other things, whether to approve Nova. But the researchers at his old lab were already reassessing their design. Like their colleagues at Rochester they had seen the evidence that shorter wavelengths reduced laser-plasma interactions. When they heard about Rochester's invention of wavelength conversion to ultraviolet using KDP crystals they realised that it would be foolhardy to build Nova without such converters, but adding them seriously inflated Nova's already substantial price tag. Foster's committee didn't play along. It recommended that DoE hold Livermore to its original cost estimate and that the lab should pay for the extra frequency converters by reducing the number of beams from twenty to ten. The goal would no longer be ignition but to refine predictions of what beam energy would be required to get to ignition.

Around this time occurred the most startling and yet most shrouded part of the inertial confinement fusion programme. The DoE set out to test the viability of inertial confinement fusion using nuclear bombs. Starting in 1978 a series of small underground nuclear tests at the Nevada test site were carried out by teams both from Livermore (which they dubbed the Halite series) and Los Alamos (known as Centurion). The aim was to use the explosions as a source of x-rays and to test DT fusion capsules of various sizes to see how much driver energy is necessary to reach energy break-even. The Halite-Centurion programme went on for ten years and its results remain classified but what information has leaked out suggests they found that 20 megajoules of x-rays are needed to reach break-even. Since hohlraums are only around 20% efficient at converting an incident laser pulse into x-rays on the target, this translates into a laser of 100 megajoules – an energy that was far beyond the state-of-the-art then and still is today. While the amount of energy seemed daunting, the results at least gave researchers confidence that ignition via laser fusion was possible.

Nova was completed in 1985 and, as the researchers had hoped, using ultraviolet instead of infrared light led to much more of the laser energy reaching the target and much less pre-heating of the fuel by hot electrons. After a few years of operation the Livermore researchers were able to achieve cleanly symmetric implosions to 100 times liquid DT density. But the energy they were getting onto the target was a far cry from Nuckolls' planned 200 kilojoules for Nova. Cutting the number of beams from twenty to ten halved that figure and the addition of wavelength converters – which were 50% efficient – halved it again to 50 kilojoules. The researchers were not even able to reach that level, however, because they had to moderate the power to prevent damaging the laser optics. So Nova was held back to powers no greater than 30 kilojoules.

Discerning readers will have noticed a pattern in Livermore's mode of operation: computer models make bold predictions about future achievements; a new laser is built; it either underperforms or the plasma physics proves more complicated than predicted, or both; models are refined and make new bold predictions; another larger laser is built; and so on. True to form, Livermore began planning for its next big machine but this time there were more than the usual number of complications. First, their computer models were much more optimistic in terms of the beam energy required to get to ignition than the results of the Halite-Centurion experiments suggested. Livermore's models predicted that energy of a few megajoules would be enough, while the nuclear tests had pointed to 100 megajoules. Even if Livermore was right, to go straight from Nova's 30 kilojoules to more than a megajoule was a huge leap for Nd:glass technology which many laser experts didn't think could be done in one jump. Part of the motivation for that big jump was that laser fusion was becoming increasingly important to weapons scientists who were constantly pushing for higher energy implosions. But big also meant expensive and the amount of money required for the next generation of machine – approaching $1 billion – was making the programme very visible to politicians in Washington, DC.

Livermore's proposal for its next laser was the Laboratory Microfusion Facility (LMF) which would have a 10 megajoule driver laser and would produce an energy output of between 100 and 1,000 megajoules – a gain of 10 to 100. At this time Rochester was also lobbying for a much more modest upgrade of its Omega laser to further demonstrate the viability of direct-drive fusion. To help decide what to do next, the National Academy of Sciences was asked to set up a panel to review the laser fusion programme in 1989 and 1990. Headed by Steven Koonin of the California Institute of Technology, the panel listened to both proponents and technological doubters and concluded that the LMF was too great a technical step and too costly to pursue at that time. Instead, the panel recommended an intermediate step, that Livermore should build a 'Nova upgrade' with a laser of a few megajoules of energy at less than half the cost of LMF. Such a facility, the panel said, would probably be able to achieve ignition and even modest gain of between 5 and 10 (down from LMF's 10 to 100) – despite the fact that there was no evidence ignition could be reached with such a beam apart from Livermore's own optimistic computer projections. The panel did, however, hedge its bets: recognising that there was no firm evidence of which target technology – direct or indirect drive – would ultimately prove more successful, it also recommended that Rochester should be funded to upgrade Omega from twenty-four beams to sixty, boosting its energy from 10 to 30 kilojoules, for the knock-down price of around $60 million.

In July 1992 Livermore came back with its proposal for the Nova upgrade. The laser would have eighteen beamlines, half in the existing Nova building and the other half in the old Shiva building next door. The light from each beamline would be split into sixteen beamlets and channelled into an upgraded Nova target chamber. The laser contained a number of technical innovations which aimed to save money as much as improve performance, such as main amplifiers that the light passes through several times, bounced back and forth by mirrors, so as to get the maximum amplification. The final beam energy would be between 1 and 2 megajoules and the facility would cost around $400 million. But political events in the wider world would soon prompt a rethink of those plans.

In the few years since Livermore had begun to plan this latest generation of lasers, the Eastern bloc had torn itself apart, the Berlin Wall had fallen and the Soviet Union itself had fractured into Russia and a handful of newly independent states. In the early 1990s the Russian economy was near to collapse and the old Cold War adversaries, worn down by decades of the nuclear arms race, were eager to talk about disarmament. One of the first items on the agenda was a comprehensive nuclear test ban, the prohibiting of all nuclear explosion tests on land, underground, in the sea, air or space. The first discussions of such a treaty began in 1991 and the following year the United States began a voluntary moratorium of testing which it has held to ever since. Without the ability to test weapons it would be impossible to design and build new ones, or even to check that existing ones were still working. The two main weapons laboratories – Los Alamos and Livermore – suddenly began to look like expensive luxuries and members of Congress started asking whether Livermore, in particular, was needed at all.

The national labs struggled to find new roles for themselves and sought to diversify into environmental research and green energy technologies. But they were given a big boost when Congress enacted a bill in 1994 that created the Stockpile Stewardship Program. The SSP was a scientific programme to understand the physics and chemistry of what happened to nuclear weapons as they aged so that measures could be taken to ensure they remained safe and reliable. This included refurbishing or remanufacturing components or whole weapons if necessary. The programme was also required to 'maintain the science and engineering institutions needed to support the nation's nuclear deterrent, now and in the future.' This meant that the country needed to keep trained weapon designers working in its national labs so that they could design new weapons if required by emerging threats in the future and, theoretically, build them without using explosive testing. Top-flight scientists were not going to sit around in the labs keeping an eye on the slowly ageing stockpile of nuclear weapons; they needed some serious science to keep them busy, so the lab directors started looking around for major new facilities that could be built as part of the programme.

The government put around $4.5 billion per year into SSP, which was a lot less than it used to spend on nuclear testing before 1992 but was still a considerable sum. The directors of the national labs held a series of meetings to decide who would get what. Sandia, it was decided, would build the Microsystems and Engineering Sciences Applications (MESA) complex where the radiation-hardened electronic components of nuclear weapons could be designed, fabricated and tested. Los Alamos got the Dual-Axis Radiographic Hydrotest Facility (DARHT) in which conventional explosives are used to compress sections of plutonium as would happen during the implosion of a nuclear bomb. The facility uses powerful pulsed x-ray beams to take ultrafast pictures of these test implosions. For Livermore the meetings decided on a laser fusion facility with which weapon scientists could study tiny thermonuclear blasts and use the data to validate computer models of nuclear weapons. The Omega upgrade wasn't sufficient for the weapons scientists' requirements and so Livermore drew up a new plan: dubbed the National Ignition Facility (NIF), it brazenly carried one of its main objectives in its name.

With the weight of the SSP behind it, NIF seemed to have an unstoppable momentum. But its huge price tag of around $1 billion and its controversial dual role – fusion energy research and stockpile stewardship – immediately drew criticism. Those in the laser fusion field, including proponents of direct drive from Rochester and the Naval Research Laboratory, argued that it was too big a technological leap and was the wrong kind of machine to be a demonstrator for fusion energy. NIF was being designed to produce laser pulses with an energy of 1.8 megajoules – sixty times Nova's energy. Laser experts feared that with that amount of energy moving through in ultra-short pulses the machine's optics would suffer frequent damage, making the laser prohibitively expensive to operate. They were sceptical that such a huge machine could produce beams of sufficient smoothness to implode targets evenly. They saw indirect drive as a dead end because hohlraums made of heavy metals such as gold – which are destroyed in each shot – would be too expensive for a working power plant. And, they said, Nd:glass was never going to cut it as a driver for a fusion energy plant because it was too inefficient at converting electrical energy into beam energy and it couldn't do the rapid-fire operation needed to make a plant commercially viable. Why then, they asked, make such a huge investment in a laser that won't take the technology of fusion energy forward?

Many in the stockpile stewardship field didn't like it either. NIF had rapidly grown to become the behemoth of the SSP and scientists at the other national labs and elsewhere attacked it for taking up so much SSP funding while having little real impact on the science of maintaining reliable nuclear weapons. There was a widespread suspicion that NIF's real role was either to surreptitiously allow the designing of new nuclear weapons or that it was just an expensive play thing to keep weapons designers busy in case they are needed in the future.

Despite the criticism, DoE pushed ahead with NIF, asking Livermore for a conceptual design. Although this isn't as detailed as an engineering design, for NIF it ran to twenty-seven volumes, a total of 7,000 pages. But the next phase, an engineering design, was put on hold in May 1994 when the then Energy Secretary, Hazel O'Leary, received a five-page memo from an anti-nuclear group called Tri-Valley Citizens Against a Radioactive Environment (Tri-Valley CAREs) which argued that any use of NIF to design new weapons could jeopardise negotiations going on at that time to renew the Nuclear Non-Proliferation Treaty. O'Leary allowed the engineering design to go ahead later in the year but she realised that NIF had better have firm scientific justification before proceeding to construction, so she asked the National Academy of Sciences to set up an inertial confinement fusion advisory committee to vet the NIF design and assess its readiness for construction. Steven Koonin of Caltech, who had chaired the 1989-90 laser fusion panel, was chosen to lead this one too.

The Natural Resources Defense Council (NRDC), an advocacy group in Washington, DC, and a vociferous opponent of NIF, accused DoE of biasing the committee. According to NRDC, several of the committee members were paid consultants to Livermore, some were awaiting decisions on bids for DoE contracts, almost all had a personal or institutional connection with DoE, and a majority had previously stated positions in favour of NIF. The committee's report, which everyone expected to approve NIF's construction, was due to be released in early March 1997, but then NRDC along with Tri-Valley CAREs and a group called the Western States Legal Foundation took the matter to court. They invoked a clause of the Federal Advisory Committee Act of 1972 which stipulates that such committees must conduct their business in public. Koonin had presided over some closed sessions of the committee, in contravention of the Act, so NRDC and the other two bodies were able to win an injunction barring DoE from making use of the report or spending any more money on it.

While deliberation of that case continued, NRDC and thirty-eight other environmental groups sought another injunction in May to prevent the start of construction of NIF, claiming that DoE had failed to comply with environmental standards when planning the facility. An official groundbreaking ceremony went ahead on 29th May although Vice President Al Gore did not attend because of the lawsuit. The official start of construction, scheduled for 5th June, was put on hold. Legal arguments continued until August when the judge in Washington's district court turned down NRDC's environmental injunction. DoE still didn't have the seal of approval of Koonin's advisory committee but it decided to go ahead without it and construction of NIF began.

This was how NIF came to be the centrepiece of the US inertial confinement fusion programme. Rochester built its upgraded Omega laser, but much of the work it did was in preparation for NIF. Some inertial confinement projects continued on a small scale at other labs – Los Alamos, Sandia, the Naval Research Lab – but these were sideshows compared to NIF. As construction began, DoE estimated NIF would cost $1.1 billion to build plus another $1 billion for operation and be finished by 2002 – like Livermore's computer simulations, this proved to be wildly optimistic.

Everything seemed to go smoothly at first, apart from a four-day hiatus in construction when crews found the 16,000-year-old bones of a mammoth on the site. A high point came when NIF's target chamber was hoisted into place in June 1999. A steel sphere 10m across and weighing 130 tonnes, the chamber required one of the world's largest cranes to lift it. The Secretary of Energy Bill Richardson was present at the ceremony that followed and he announced that the project was 'on cost and on schedule.' He said the same thing in Congress when questioned about progress. Then everything seemed to go wrong at once.

First Michael Campbell, the long-time Livermore physicist who was leading the project, stepped down in August after an anonymous whistleblower told Livermore managers that he had never finished the PhD from Princeton University that he claimed to have. Then only days later it emerged that the project was suffering from numerous technical problems and was, in fact, a year behind schedule and $200 million over budget. Worse still, the project's leaders had hidden this information from Richardson, which had led him to give incorrect information to Congress. A number of staff were fired or demoted and there were financial penalties for the Livermore director and the University of California, which manages the lab. Numerous investigations were launched to find out what had gone wrong, including one by the influential Government Accountability Office, the investigative arm of Congress. A Congressional appropriations committee directed DoE to develop a new schedule and cost estimate for NIF and to present it to Congress by 1st June, 2000 or prepare to terminate the project. Livermore's freedom to manage the project as it saw fit with little scrutiny – a side-effect of its role as a weapons lab – was at an end as NIF became one of the most closely monitored projects in the history of science.

NIF's technical troubles had started hundreds of miles away at Sandia National Laboratory in New Mexico. With its expertise in pulsed power experiments, Sandia was contracted to produce 200 huge capacitors – stores of electric charge – which would be used to power the flashlamps that pump NIF's laser amplifiers. But during tests the material of the capacitor was vaporising and the pressure inside the device caused its metal cladding to fly off like shrapnel. The capacitors had to be redesigned with a centimetre-thick steel shell on the outside and pressure-escape doors at the bottom. The companies that were making the neodymium-doped glass for the amplifiers had trouble keeping out impurities, which led to more delays. Another constant bugbear was dust: if there was any dust on the surfaces of the optics when beams passed through it was likely to ignite and damage the surface. But keeping the levels of dust down in a building the size of a football stadium proved to be a complete nightmare. The building itself turned out to be a problem because it was built before the design of the laser system had been finalised, leading to a tight fit that left little room for maintenance.

DoE reported back to Congress on 1st June, 2000 with an 'interim report' that did not detail cost or schedules, but it soon emerged that it was estimating a final cost of $3.3 billion and completion in 2008. The Government Accountability Office submitted its report in August, concluding that 'NIF's cost increases and schedule delays were caused by a combination of poor Lawrence Livermore management and inadequate DoE oversight.' Its estimate of NIF's total cost was nearer to $4 billion. Energy Secretary Richardson said that he didn't want to ask Congress for more money for NIF so would find the cash from elsewhere. Fearing that some of their facilities might be cut back to pay for NIF, the other weapons labs suggested cutting back on the number of NIF beamlines to save money. But somehow the wounded NIF survived. It limped on, heavily scrutinised, but without much further incident, until 2009 when it was declared complete, seven years late and having cost almost twice the original estimate.

Accident-prone it may have been, but NIF's many opponents failed to stop it being built or to scale it back. With the backing of weapons designers and a measure of institutional inertia, it made it to the finish line. All that remained now for the world's biggest laser was to see if it could achieve ignition.

**One of two laser bays at the National Ignition Facility. Here beams are amplified in power before converging on a fusion target.** (Courtesy of Lawrence Livermore National Laboratory)

CHAPTER 7

# One Big Machine

IN MOST SCIENCES YOU BUILD MACHINES TO ALLOW YOU TO DO experiments. In fusion, the machine is the experiment. You build it to see if it will work and how it works. Because fusion machines take such a long time to build, fusion scientists are always looking one, two or more machines into the future: while they're building one, they're always planning more. So it was in the late 1970s, when the construction of JET and TFTR was only just beginning, that many researchers were beginning to think about what was to come next.

The route to fusion energy that had been mapped out earlier in the decade had three stages. First was a demonstration of the scientific feasibility of fusion, in other words a gain greater than 1, and the big tokamaks being built in Princeton, Culham and Naka were expected to take care of that. Next would be technical feasibility, a machine that would produce large amounts of energy while testing some of the technologies that would be needed in a fusion power station, such as superconducting magnets, a system for extracting heat to raise steam, and a method for breeding tritium for fuel. The final stage was commercial feasibility – a prototype power reactor.

So while the three big machines still existed only on paper, some of the more far-sighted planners were already thinking about the even bigger machine, the 'engineering reactor' that would come after. One of them was Russian theorist Evgeniy Velikhov. Velikhov was a rising star in the Kurchatov Institute's fusion department in the 1960s where he teamed up with fellow young theorists Roald Sagdeev and Aleksandr Vedenov – soon dubbed by colleagues as the 'holy trinity.' Plasma theory was too narrow a field to contain his talents and he later branched into lasers as well as computers and automation. He was also a skilful political operator – he knew how to play the Soviet system of patronage and political influence. His star was rising so fast that in 1973, at the age of just 38, he took over the reins of Russia's fusion programme following the death of Lev Artsimovich. In 1974 he was made a full member of the Soviet Academy of Sciences and three years later was elected its vice president.

Before his death, Artsimovich sent Velikhov to represent the Kurchatov Institute in discussions at the International Atomic Energy Agency (IAEA) in Vienna. Ever since the 1958 Geneva conference, fusion scientists had maintained a constant dialogue between East and West. The IAEA organised its regular fusion conferences which all nations could attend. There were visits to each other's labs and exchanges of information but relations stopped short of any formal cross-border collaboration apart from that of Euratom. That started to change when Velikhov, along with Amasa Bishop of the US Atomic Energy Commission and IAEA chief Sigvard Eklund, formed the International Fusion Research Council (IFRC) in 1971 to advise the agency in its efforts to coordinate worldwide fusion research. Although the IFRC was simply a group of advisers, Velikhov hoped it would prove influential in moving fusion research towards closer collaboration. He already suspected that an engineering reactor might be so large as to be beyond the capabilities of a single country's fusion research programme.

Velikhov was not the only one thinking along these lines. Within all the fusion programmes, researchers were beginning to realise that as soon as they had cracked the problem of getting a plasma to burn they would have to learn how to handle the neutrons, extract heat and breed tritium. One of those was David Rose, an engineer who was hired by the Massachusetts Institute of Technology in 1958 when it set up its Department of Nuclear Engineering. In the late 1960s he carried out a detailed study of how energy in a fusion reactor would be exchanged between different types of particles – electrons, deuterium and tritium ions, and alpha particles – and how you would inject fuel into the plasma and remove helium exhaust. His calculations suggested that a fusion reactor would be economically viable, but it would need to be big. In 1969 he co-organised a meeting at Culham, the first meeting to consider the engineering issues of a fusion reactor, and this encouraged many more engineers to get involved.

As the 1970s progressed and tokamaks grew bigger and performed better, the need for engineering solutions became more pressing. In 1977 Rose invited senior engineers from different countries to a meeting to discuss how they might better work together. The group weren't sure how to form an international collaboration but concluded it should probably be organised by the IAEA. It turned out the IAEA was already moving in that direction. The agency's chief, Eklund, asked the IFRC for suggestions of how the IAEA could take a more active role in fusion research and Velikhov quickly stepped up with a plan he had already worked out: an international collaboration to design a reactor with the express purpose of testing the technology necessary for a commercial reactor.

The project, which was named the International Tokamak Reactor or INTOR, began in 1979. Each of the four participants – Euratom, Japan, the Soviet Union and the United States – nominated four researchers who would meet in Vienna several times a year for workshops that lasted between four and six weeks. When the researchers returned home they would delegate work to their colleagues to be carried out before the next workshop and so the network of researchers involved became very broad.

The majority of fusion researchers didn't take INTOR too seriously; they were too busy getting TFTR, JET, T-15 and JT-60 up and running. The regular trips that a few of them took to Vienna were just a sideshow, but the INTOR workshops did gradually build up a database of knowledge on how fusion reactors work, taking results from all the fusion programmes. The workshops produced a number of reports describing the theoretical reactor they were working towards, each report increasing in detail and sophistication. But perhaps INTOR's main achievement was that it showed the very different traditions and methods of the various fusion research programmes could work together and get things done.

But by the mid 1980s it was clear that INTOR wasn't going anywhere. Although its designs for an engineering reactor were highly praised by researchers, there just wasn't any political support for an international project to build a giant fusion reactor. Velikhov was frustrated that his ambition to move quickly towards fusion power generation had stalled. The Soviet Union was certainly in no position at that time to build an engineering reactor itself: the economy was in bad shape and work on the T-15 reactor had all but come to a standstill. Velikhov needed a way to get political support behind INTOR and, by an incredible stroke of luck, an opportunity fell in his lap to take his appeal right to the top. That opportunity came in the shape of Mikhail Gorbachev, an old university friend of Velikhov's, who in March 1985 became general secretary of the Community Party of the Soviet Union – Russia's *de facto* leader.

The two had been at Moscow State University at the same time, Velikhov studying physics and Gorbachev law. Gorbachev became active in the Communist Party and on leaving university rose swiftly through its ranks. In the early 1980s, the deaths in fairly rapid succession of Soviet leaders Leonid Brezhnev, Yuri Andropov and Konstantin Chernenko led the Politburo to decide that younger leadership was needed. So, just three hours after Chernenko's death, the Politburo elected its youngest member – Gorbachev, aged 54 – to the top job. He immediately made his mark as a reformer with his policies of *glasnost* (openness) and *perestroika* (restructuring) which sought to loosen the shackles of the old regime. In foreign policy he made decisive moves to reduce East-West tension. He withdrew SS-20 intermediate-range nuclear missiles from central Europe less than a month after taking office and within six months proposed that the Soviet Union and US both cut their nuclear arsenals by half.

As head of the country's fusion programme and an old friend, Velikhov met with Gorbachev and described to him how an international project to build a fusion reactor along the lines of INTOR, with Soviet and American researchers working side-by-side, could help to diffuse Cold War antagonism. Gorbachev eagerly adopted the idea and in his first foreign trip, to France in October, he discussed the idea with President François Mitterand and received a positive response. Gorbachev's next foray abroad was to meet US president Ronald Reagan in Geneva at their first summit. In the weeks running up to this November meeting Velikhov worked feverishly with contacts in the White House to get something ready for the leaders to agree on, despite strenuous opposition from the Pentagon which was concerned about valuable software and technology being handed over to the Soviet Union. The often tense and argumentative summit meeting was dominated by discussions of human rights and the Strategic Defense Initiative, Reagan's proposed nuclear missile shield. No break-

**Ronald Reagan and Mikhail Gorbachev at the Geneva summit, November 1985.** (Courtesy of ITER Organisation)

throughs were made in East-West relations but the two leaders did establish a relationship that would stand them in good stead in the future. The rather bland final communiqué did however include – as one of its twelve bullet points – a commitment by the two powers to work together and with others to establish the feasibility of fusion energy 'for the benefit of all mankind.'

The government agencies responsible for fusion in each country were slow to get moving in forming a collaboration. It was only after the Reykjavik summit between Reagan and Gorbachev in October 1986, when the matter was brought up again, that the bureaucrats got moving. The United States and Soviet Union, together with Euratom and Japan, formed a Quadripartite Initiative Committee to discuss the idea but talks went far from smoothly. Euratom was already well advanced in planning its follow-up to JET, the Next European Torus, and Japan had its own plans too, so neither of these saw the urgent need for another reactor design. US defence officials were still concerned about the transfer of sensitive technology and the Russians wanted any joint design work to be done in a neutral country. After somewhat tortuous negotiations the four partners agreed to work together for a couple of years to produce only a conceptual design – a broad outline that doesn't get into the detail needed for actual construction. Paulo Fasella, Europe's representative at the negotiations, gave the project a name: the International Thermonuclear Experimental Reactor, or ITER. Fasella, a highly educated man who had a glittering career in biomedicine before joining the European bureaucracy in Brussels, pointed out that *iter* in Latin means 'the way.'

The headquarters for the conceptual design work was Garching in Germany and the project proceeded in a similar vein to INTOR: each partner seconded around ten researchers to the project who would spend several months each year in Garching and then delegate work to their colleagues back home. The collaboration of researchers from four different traditions didn't always go smoothly. One American researcher described it like this: the Europeans would rant and rave passionately about every issue and the Americans had to explain to the Japanese that they *didn't* really mean it; the Japanese would explain their point of view very calmly and quietly and the Americans had to explain to the Europeans that they *did* really mean it. There were scientific differences too. The Japanese were keen to have a reactor that could demonstrate continuous, or steady-state, operation, while the Europeans wanted the highest possible gain.

In spite of the differences, this project had a different feel to it than INTOR. Because the scientists were now working as directed by their political leaders it somehow felt more real, as if this reactor would actually get built. That dose of realism was also reflected in the design of the machine. When predictions are uncertain, fusion reactor designers tend towards bigger and more powerful machines in the hope that more plasma and stronger fields will swamp any inadequacies. So while INTOR had predicted that a tokamak 12.4m across with a plasma current of 8 MA would be enough to reach ignition, the ITER conceptual design called for 16.3m and 22 MA.

After two years of work the ITER team had come up with a conceptual design that they could all more or less agree on. The question then was, what to do next? The original plan had been to move straight on to drawing up an engineering design – an exact set of plans ready for construction. But the world had changed since ITER had been dreamt up as a way of alleviating Cold War tensions. The Iron Curtain was falling apart and even the Soviet Union itself would soon cease to exist. ITER's political *raison d'être* had evaporated and neither was there an economic need for it: energy was not high on the political agenda in the early 1990s. The project's momentum carried it forward, but it took the four partners two years to decide on how to proceed. A major sticking point was where the design team would be situated. The dying Soviet state and then the new Russian Federation were in such dire economic circumstances that they could make little real contribution to the project apart from scientific brainpower, but none of the other three partners were ready to concede the design team to the other two. They came up with the unwieldy compromise of splitting the team in three: one part in Garching would be responsible for the plasma vessel and everything inside it; another group in Naka would take on everything outside the plasma vessel – including superconducting magnets, power supplies and buildings – and the final group in San Diego would be in charge of overall integration, physics and safety.

By 1992 the engineering design project was ready to go. The teams were given six years to develop the final design. All that was still needed was an overall project leader. There were few people in the world who had the right mix of experience in engineering, plasma physics and the management of large projects. But there was one obvious choice, the person who had for nearly two decades led the design and construction of JET and guided it through the demonstration of H-mode to the first D-T burning in 1991: Paul Henri Rebut. And so Rebut left his beloved JET and moved to San Diego to take the reins of the ITER megaproject.

The designers of ITER were faced with something of a dilemma. This reactor was meant to be a demonstration of technical feasibility but the previous stage in the three-step progress towards fusion power – scientific feasibility – had not quite been achieved. TFTR only reached a gain of 0.3 in 1993 and JET would get to 0.7 in 1997. Only JT-60 managed a gain greater than 1, but that was equivalent gain – what it would have been if tritium had been used. As a result, ITER was saddled with the twin goals of demonstrating scientific feasibility and testing the technologies needed for a power reactor. The problem was that these two goals are not easily fulfilled by a single reactor design.

For an engineering reactor you want a plasma that is stable and quiescent, that will burn for long periods at high gain – mimicking how a working reactor would behave. Such a plasma would be a tool for engineers to test things such as the best lining for the reactor vessel, known as the first wall, so that it will not pollute the plasma and will stand up to years of neutron bombardment. Engineers would also want superconducting magnets because they would reduce the energy demands of a power-producing reactor. They would also want to test blanket modules – sections of the vessel wall – containing lithium which would be converted into tritium fuel by the bombardment of neutrons from the fusion reactions.

But if you haven't yet achieved gain greater than one, you would ideally want a very different reactor – something that is more like an experimental apparatus than an industrial prototype. You wouldn't want a reactor with a fixed configuration because you need to try out all possible permutations to get the highest gain. You wouldn't necessarily want a quiescent plasma; you would want to be able to push it to the edge of stability in pursuit of the best performance. And you certainly would not want complications such as superconducting magnets and tritium breeding blankets that make it harder to interpret results. So, ITER was destined to end up as something of a compromise.

When Rebut arrived in San Diego to take control of the project, he hit the ground running and immediately began remodelling ITER in the image of JET. This is not entirely surprising since JET was the largest tokamak around and had proved very successful. Rebut's team drew up a design with a D-shaped plasma, similar to JET's, and with a divertor around the bottom of the vessel, just like the one that at that time was being fitted to JET. The only fundamental difference was the superconducting magnets. Under Rebut's leadership the already large conceptual design grew even bigger, to a machine nearly 22m across. He increased the number of superconducting magnets to hold this huge volume of plasma in place. The whole magnet system – twenty toroidal magnets and nine poloidal magnets – would weigh a total of 25,000 tonnes, roughly the same as the Statue of Liberty.

Rebut didn't want ITER to use superconductors. It was possible to achieve high gain with conventional copper magnets but the partners wanted technology that was 'reactor-relevant.' Rebut argued that superconductors just made everything more complicated. Superconducting magnets are much harder to make than copper ones and they must be enclosed in a secure container called a cryostat so that they can be surrounded in liquid helium to keep them chilled to close to absolute zero (around -270°C). Because ITER will be producing large quantities of heat from its fusion reactions, the magnets must be shielded or they will warm up and stop operating. So the inside surface of the vacuum vessel would be lined with replaceable steel panels cooled by water flowing inside them.

ITER's divertor was another key piece of technology because it was the only solid object in direct contact with the plasma during normal operation. Its roles, extracting the exhaust gas from fusion (helium) and absorbing heat, required a material with a high melting point that is able to withstand prolonged particle bombardment. ITER researchers made use of many tokamaks with divertors around the world in their search for the right material. Rebut considered some quite radical solutions to ITER's many engineering challenges, such as using highly reactive and flammable liquid lithium as a coolant for some reactor components, on top of its role as material to breed tritium for fuel.

The first six months in charge was a gruelling time for Rebut. Every weekend he would fly one-third of the way around the world to the next worksite, spend a week there, then move on again. Despite this itinerant lifestyle he found it difficult to keep up the communication between the three sites. It was a very different operation from JET where he had his whole team around him. Though a brilliant engineer, Rebut was not a natural manager; he found it hard to delegate and so took on much of the design work himself. The international partners didn't like his style of leadership. They wanted their own researchers, in Naka and in

Garching, to play a greater role in the design. ITER was too big a project for it to become a Rebut one-man-show and in San Diego he didn't have Palumbo and Wüster to shield him from the politicians. The ITER partners wanted a more collaborative approach so after just two years in charge Rebut was eased out.

It was no easier finding a suitably qualified leader than it had been two years earlier, but the person the partners decided upon was not just another European but also another Frenchman. Robert Aymar had led the building of France's Tore Supra, the follow-on to its Tokamak de Fontenay aux Roses (TFR). But when they asked Aymar if he would lead the ITER project, he said no.

A contemporary of Rebut, Aymar had spent most of his career working at France's Commissariat d'Énergie Atomique (CEA) on plasma physics. He was inspired by attending the 1958 Geneva conference to pursue fusion because of its possible value to society. By the late 1970s he was in charge of France's fusion programme and when researchers had done as much as they could with TFR he set out to build Tore Supra. This reactor would be the first large tokamak to use superconducting magnets. With the strong steady magnetic fields they produced, Tore Supra would be able to hold its superheated plasma for minutes at a time instead of the seconds possible in a conventional tokamak.

Aymar realised that for such an ambitious project to be successful he needed to have all of the team together in one place, not scattered around universities and labs across France. Just as was happening at JET across the channel, he needed a team dedicated to this one goal and nothing else. During 1984 and '85 he persuaded some 300 families to move down to Provence as Tore Supra took shape at a CEA lab complex in Cadarache. Aymar and his team completed Tore Supra in 1988 and in the process created a powerhouse of fusion research at Cadarache. Recognising that success, the CEA offered him the job of heading its whole basic physics division. Aymar was in his element: with a staff of 3,000 scientists he was now responsible for the commission's work in many areas beyond plasma physics, including nuclear and particle physics. But then ITER came calling.

At that time, in 1994, the pinnacle of fusion achievement was TFTR's D-T shots with a gain of around 0.3 – hardly a convincing demonstration of fusion. In a few years' time JET would get closer to break-even but to Aymar's eyes fusion research had a way to go before it could confidently demonstrate high gain. That being the case, the machine that Rebut had been designing in San Diego was the wrong sort of machine. It was too much like an engineering reactor and didn't have the flexibility that may prove necessary to achieve the physics goal of high gain.

Despite his earlier refusal, ITER's backers contacted Aymar again and asked him to reconsider. Aymar thought hard about it. Although he was enjoying himself running the wide array of fundamental physics at the CEA and was concerned about the current ITER design, perhaps it was his mission in life to guide fusion a bit further along the road to real power generation. He accepted the job, but he wasn't happy about it. As soon as he was on board he set off to visit the three ITER work sites – Naka, Garching and San Diego. His aim was to steady nerves and build confidence among the researchers after the change in leadership.

But holding together an unwieldy design project would soon prove to be the least of Aymar's problems as, in autumn that year, the Republican Party took control of both houses of the US Congress. Two months after the Congressional elections, in January 1995, in a conference centre just outside a wintry Washington, DC, Anne Davies, then head of the Department of Energy's fusion programme, told the directors of America's fusion laboratories to begin preparing for the worst.

At that time, DoE was spending $350 million per year on

magnetic fusion, but once construction began on ITER and Princeton's proposed Tokamak Physics Experiment (TPX) much more money would be needed. Building ITER alone would consume more than the whole of the current budget. America's fusion leaders didn't have to wait long for the axe to fall. Later in the year, when Congress set the 1996 budget, it awarded magnetic fusion just $244 million. This was barely enough to keep America's domestic fusion programme going, let alone pay for an expensive new reactor to be built somewhere overseas.

Anne Davies and her DoE staff now had the unenviable task of deciding which programmes would survive and which would not. Lab directors and university department heads started jockeying for position to ensure that their reactor or research project was not cut. ITER, as the most expensive item on the wish list, started to attract the wrong sort of attention. Some in the US fusion community agreed with Aymar and others that the leap to an engineering reactor was too risky and believed something more modest should be tried first. With limited resources, some argued, why should the US be supporting a hugely expensive machine in another country – whose success wasn't guaranteed – while denying fusion scientists at home money to carry out any meaningful research? ITER also had few friends in Congress. It was typical for a large project to be championed by the senator or member of Congress representing the state or district where the project will be built. ITER didn't yet have a site, and it almost certainly wouldn't be in the United States, so it lacked such a champion.

But funding and politics weren't ITER's only problems. In the mid 1990s two researchers from the Institute for Fusion Studies at the University of Texas in Austin, William Dorland and Michael Kotschenreuther, developed a new computer simulation of the plasma inside a tokamak and it produced some very unwelcome news for ITER. The very hottest place in the plasma – where fusion is most likely to happen – is the centre, with the surrounding plasma acting as a layer of insulation slowing down escaping heat. That's one of the reasons why a bigger tokamak is better than a small one, because there is more insulating plasma around the hot core. Turbulence is the enemy of this insulating effect because it mixes plasma from the hot core with cooler outer layers, helping heat to escape towards the edge. Tokamak designers knew about this turbulence effect and used scaling laws to extrapolate from known amounts of turbulence in existing reactors to predict how much turbulence there would be in future reactors such as ITER.

In contrast Dorland and Kotschenreuther predicted in a detailed way how plasma would behave in tokamaks of different sizes and under different conditions. They verified their simulation by adjusting it to mimic existing tokamaks and it was able to predict how they behaved pretty well. When they presented their simulation and its predictive power at conferences, other researchers were impressed. Then the pair applied their model to the proposed design for ITER and received a shock. The simulation predicted that in the large volume of plasma in ITER there would be a lot of turbulence, more than was predicted by scaling laws. This turbulence would bleed heat away from the plasma core to such an extent that, Dorland and Kotschenreuther estimated, ITER might not be able to achieve temperatures necessary for fusion.

When they presented these new results the reception was far from warm. ITER was a multi-billion dollar project to which many researchers had devoted their whole working lives; they were not going to see it sunk by a pair of computer geeks. The simulation was now subjected to much more intense scrutiny and its authors to harsh criticism. The Texas pair stuck to their guns and to this day fusion researchers remain divided over whether their predictions are correct – a working ITER will be the final proof. The research did not derail ITER but it did provide valuable ammunition to the project's critics.

* * *

By 1997 the ITER team was putting the finishing touches to the reactor's final design, described in a huge 1,500-page report. The machine it described remained largely along the lines laid out by Rebut: the plasma vessel was 22m across – more than two and a half times the size of JET – with enormous, and costly, superconducting magnets to hold the plasma in place. The price tag of $10 billion was equally awe-inspiring. The target was to steadily produce 1.5 gigawatts of heat – one hundred times JET's record-breaking output in 1997. And while JET required 14 MW of external heating to keep the reaction running, the ITER design called for 150 MW of heat, from both neutral particle beams and radio-frequency heating. The plan was for ITER to also achieve ignition – heating itself with the energy from alpha particles created by fusion reactions. At that time, no reactor had even got close to ignition.

As the scale and cost of the proposed ITER started to become apparent, US participation in the project was more and more difficult. Ever since the election of the Republican Congress in 1994 the Department of Energy's fusion budget had been repeatedly chipped away, leading to the cancellation of TPX and the closure of TFTR. Republican representative Jim Sensenbrenner, a fusion sceptic, was appointed chair of the House Committee on Science and Technology, which was ultimately responsible for US fusion funding. Fearing the demise of fusion research in the US, DoE officials were applying tremendous pressure on Aymar and his team to keep the cost of ITER down. Aymar himself was beginning to have doubts about the design, that it was too big, too expensive and too great a leap from current knowledge. On the quiet, Aymar asked researchers at the Garching site to start work on a more modest design, one that would still generate a lot of power and so show that fusion energy was feasible, but also one that was an evolution from existing tokamaks rather than a revolution.

Others were thinking along similar lines. With only $50 million a year to spend on ITER, Davies at the DoE knew that the proposed machine was far beyond America's means. DoE officials suggested scaling back the ITER design to build a smaller reactor, with less ambitious goals and a price tag cut in half – a plan dubbed 'ITER Lite.' Some US researchers, aware of the fact that funding ITER would probably squeeze fusion research spending inside the US down to nothing, were proposing an even more radical retreat: abandoning the idea of a single huge machine altogether and instead upgrading existing tokamaks and carrying out an international campaign of research to better understand burning plasmas. With doubts about turbulent heat transport in the back of their minds, they argued that if there is that much uncertainty it's probably best not to build a huge, expensive reactor that could still fail. At academic conferences, among US researchers at least, the talk in the corridors suggested that ITER was all but dead.

According to the schedule, now that the design was complete, the partners should choose a site for the reactor in 1998 and then start building, aiming to complete ITER in 2008. But it soon became apparent that the US was not the only partner having money troubles. Japan had long been one of ITER's most enthusiastic participants and many expected the machine to be built there. But in the spring of 1997, Japan no longer had the money to match its enthusiasm. At that time Japan's postwar economic miracle had ground to a halt. Often referred to now as Japan's 'lost decade,' the 1990s saw the country try to spend its way out of recession with ambitious public works. When this failed to get the economy moving, the government was forced to tackle the budget deficit by cutting spending. Along with other big-ticket science projects, Japan was forced to cut back on ITER, and so it asked for a three-year delay in the start of ITER's construction.

For Jim Sensenbrenner and the House science and technology committee this just reinforced their view that the ITER project was in terminal decline. According to the committee ITER, at $10 billion, was too expensive; it was questionable whether it would work (c.f. Dorland and Kotschenreuther); and it couldn't even be considered a viable project since it didn't have a site. It also made the committee uncomfortable to be giving so much taxpayers' money to a project that was not controlled by the US. The committee allowed US membership of the engineering design collaboration to run its course, but from 21st July, 1998 US participation in the project was stopped. Scientists working in Garching and Naka returned to the US. The worksite in San Diego was closed down and non-American researchers sent home. US researchers were forbidden from participating in ITER activities or meetings, even as observers. The unprecedented East-West and then global cooperation that had existed in fusion research for the four decades since the 1958 Geneva conference, and that had outlasted the Cold War, was terminated by order of the US Congress.

The ITER project was in a critical condition. With Europe now the only one of the three remaining partners that still had a sizable fusion programme it was hard to see a way forward. Some in Europe argued that they should just abandon ITER and make something more affordable, a European successor to JET. In Japan, there was a crisis of confidence. Ever since the Second World War, Japan had taken a lead from the US in matters of foreign policy. With America out of the picture, could they trust the Europeans?

The partners set up a working group to consider two options: a large ITER-like machine able to study both the science and engineering of a burning plasma, or a number of smaller machines that could address different issues. The group concluded that the only way to examine how all the many questions surrounding burning plasma interrelate was in a single integrated machine with long pulses and alpha particles as the dominant source of heat. The Japanese decided to stay on board, so the teams in Garching and Naka were set the task of designing a new, smaller machine, costing half as much as the 1998 design but retaining as many of ITER's technical goals as possible.

Aymar's hunch that a slimmed-down ITER might be needed proved prescient and much of the necessary redesign work had already been done at Garching. In 2001 the teams presented a new final design for a reactor with a vessel that was 16.4m across, instead of the original 22m, and capable of carrying a plasma current of 15 MA, down from 21 MA. The output power, at 500 MW, was also one-third the previous goal, but the biggest sacrifice was that the new ITER was no longer expected to achieve ignition. Instead of running on alpha particle heating alone, the reactor was likely to need at least 50 MW of external heating to top up the alphas and keep the plasma burn ticking along. That still meant a gain (*Q*) of 10 but, yet again, one of the major milestones of fusion energy seemed to be retreating out of reach.

The redesign did reduce the cost of the project to a slightly less eye-watering ⇔5 billion, but now the partners had to come to terms with the reality of choosing a site and building it. While the design of the reactor had been left for the scientists to decide – within the limits of the budget available – the choice of site would be a largely political decision and all the researchers could do was sit tight and hope for the best. The plan was for the nation that was chosen as host to shoulder the greatest share of the construction cost – because of the economic benefits of having the reactor on its territory – with the remainder divided equally among the other partners. But only a small part of each partner's contribution would be in the form of cash paid over to the yet-to-be-created ITER organisation. Most of it would be contributions 'in kind': components for the reactor that would be manufactured by each partner's home industries and then shipped to the site. ITER managers would carefully divide up and parcel out the construction work so that each partner made an appropriately sized contribution while its industry learned the skills that will be needed to build future commercial fusion power stations. Everyone wanted a share of the knowledge that could turn into a multibillion dollar industry. But the question remained: who would play host and would it be a welcome boon or a crippling burden?

The surprise first entrant into the site contest in June 2001 was Canada, which was not at the time a member of the ITER project. The offer was being promoted by a consortium of companies led by the power utility Ontario Hydro. It had a site on the north shore of Lake Ontario just east of Toronto, next to the Darlington nuclear power station, which was already licensed for construction of a nuclear plant. In many ways, the offer made sense: Canada has plentiful supplies of tritium fuel because it is a by-product of its home-grown Candu fission reactors; Ontario Hydro would benefit by supplying electricity to the project; and, situated halfway between Europe and Japan, the Darlington site presented a compromise solution that might also lure Canada's southern neighbour to rejoin the project.

But other partners were not ready for Canada to waltz in and carry off the prize. In Europe, Germany had always been the most enthusiastic supporter of ITER and had considered offering to host it, but the cost of reunification with East Germany after the fall of the Berlin Wall meant that it was now not so keen. France, however, had an almost ready-made site: Cadarache, the Commissariat d'Énergie Atomique lab that Aymar had built up into a fusion powerhouse. It had available land and supplies of power and cooling water already built for Tore Supra, plus support from the French national and regional governments. Japan was weighing up three potential sites: Tomakomai on the northern island of Hokkaido, Rokkasho at the northern tip of the main island Honshu and Naka, north of Tokyo, home to JT-60. Russia's economic malaise still ruled it out as a potential host.

Back in the United States, fusion researchers and the DoE were busy trying to figure out what to do next. Many wanted to get back into ITER as soon as possible and they were encouraged by the fact that the remaining ITER members were working on a smaller and cheaper design. But for a while at least, rejoining the project was not politically possible. The community began working on a design for a new home-grown reactor. Known as the Fusion Ignition Research Experiment (FIRE), the reactor would study the physics of ignition and was championed by Dale Meade, the tall and affable former deputy director of the Princeton fusion lab. Meanwhile Bruno Coppi of the Massachusetts Institute of Technology put forward another alternative: a reactor called Ignitor which followed the model of MIT's Alcator tokamaks in using very high magnetic fields to get strong confinement and heating. His Ignitor proposal would go all-out to show that ignition was physically possible, but others considered it would do little else to aid progress towards a power-producing reactor.

In July 2002 a few dozen senior figures from the US fusion community gathered at the Snowmass ski resort in Colorado for a two-week-long conference to consider what they should do next. Ultimately it was up to the DoE and Congress to decide what to back, but the scientists knew that if they presented a united front they were more likely to get what they wanted. There were three options on the table: rejoin ITER, build FIRE or build Ignitor. There were vigorous debates, interspersed with walks in the mountains, but a poll at the end showed where the researchers' hearts really lay: they voted 43-to-1 in favour of rejoining ITER, with FIRE held in reserve if that proved impossible – even Meade voted for ITER. An advisory panel to the DoE considered the same question later that year and all members of the panel voted for ITER with FIRE as a backup, apart from Bruno Coppi who voted for Ignitor.

There remained the problem of persuading politicians that the project they so decisively rejected a few years earlier was now researchers' top priority, but that now didn't seem as impossible as it once would have. The Republican Party was not the dominant force in Congress that it had been and then there were the terrorist attacks on 11th September, 2001 which caused a change in the tenor of American politics. George W. Bush had replaced Bill Clinton in the White House in January of that year and his administration did not seem any more enthusiastic about fusion than Clinton's. The highlight of his early energy policy seemed to be a proposal to open up the Arctic National Wildlife Refuge to oil drilling companies. In the aftermath of 11th September, everything changed: suddenly energy security was high on the agenda. How could the US ensure its energy supply in the event of another large terrorist attack or conflict in the Middle East? Research into energy technologies that didn't rely on imports of fossil fuels were suddenly flavour of the month.

Just six months after the attacks, with fusion scientists now unambiguously backing ITER, government officials began to look into how the US could rejoin the project, reportedly at the suggestion of President Bush himself. In his state-of-the-union address in January 2003, the President made a commitment to develop cleaner energy technologies and generate more of it at home rather than importing oil. Two weeks later a US delegation travelled to St. Petersburg for an ITER council meeting to begin the process of rejoining. The US was not alone in its new enthusiasm for ITER: China and South Korea also joined the collaboration in 2003.

In the space of five years the fortunes of the project had made a dramatic turnaround. In 1998 ITER had been on the brink of collapse. Now it had six partners – one of whom encompassed most of Europe – several sites vying to host it and a design that everyone believed in. At this high point, Robert Aymar decided it was time to hand the leadership over to someone else. He had supervised the completion of two designs and navigated ITER through its greatest crisis: now another project in trouble had come calling. At CERN, the European particle physics lab near Geneva, construction of its giant particle-smasher, the Large Hadron Collider, was well over budget and struggling to stay on schedule. So they called Aymar. He said at the time that he was too old to embark on a job like constructing ITER, which would go on for another ten years.

Once the question of the site was settled, the ITER project would be transformed into a fully fledged international organisation charged with building the reactor. That would mean new leadership, so in the meantime Aymar's deputy, the Japanese plasma physicist Yasuo Shimomura, was made interim director. The whole ITER operation was in a state of suspended animation as the team waited for politicians to decide where it would be built. Japan whittled its roster of proposed sites down to one, Rokkasho, while Europe had acquired a second one, Vandellos near Barcelona in Spain. The ITER council, a biannual meeting of delegates from the partner governments, looked into the merits of the four sites and declared them all suitable from a technical point of view. If all of them would work, how were they to choose? The backers of each site began pushing other attributes, as if selling package holidays, to try to persuade other partners: Rokkasho will have Western-style housing for staff and an international school for their children; Cadarache has the weather and ambiance of Provence and the nearby Cote d'Azur; while Darlington is a stone's throw from the cosmopolitan city of Toronto. A date was set for the council meeting that would make the decision: December 2003 in Washington, DC.

The EU decided that it had to choose between its two candidate sites – Cadarache and Vandellos – to increase its chance of success in Washington. Technical assessments concluded that either would work. Building in Spain would be cheaper but the site, next to an existing nuclear power station, didn't have any science institutes nearby. At Cadarache there was already a wealth of scientists and engineers on hand if help was needed, but the site was far from the sea so the transport of large and heavy components could be tricky. Debate over the two sites raged on through the summer and autumn. Favours were called in; backs were scratched; and different countries lined up behind their favourite sites. Although the process was divisive at the time, it helped to cement Europe's determination to win ITER. In a sense, the Europeans felt that they had earned it. While the US fusion budget was slowly whittled away and the Russian and Japanese programmes were undermined by their struggling economies, Euratom had kept the whole project afloat, especially during the years after America's withdrawal. Hosting ITER on European soil would be the payback.

The person who found himself in charge of Europe's effort to win ITER was Achilleas Mitsos, a gruff economist from Greece and a specialist in European integration who joined the European Commission, the EU's civil service, in 1985. As is the custom at the Commission, every few years he was moved to a different job, managing such issues as social and economic cohesion, education and training, and socio-economic research, before eventually becoming director-general for research in 2000. By now a seasoned operator in the Brussels bureaucracy, Mitsos was unruffled by the tussle between Cadarache and Vandellos. It had required some diplomacy: to appease Spain when Cadarache was chosen, it was promised that the organisation that would eventually be needed to manage Europe's part of ITER construction would be based there. But that did little to prepare him for what was going to come next.

The Washington meeting of the ITER council was to be the

project's turning point, the moment it changed from an idea on paper into an international collaboration intent on building a fusion reactor. President Bush was on standby to come in and lend authority when it came time to sign an agreement. Everyone expected a deal to be done, but the events of 9/11 cast a dark shadow over the meeting. The Iraq war had begun only nine months earlier and relations between the United States and France were in a deep freeze because of French opposition to the war. Now the EU had the temerity to come to the negotiating table with a plan to site ITER in France.

Tensions began to simmer before the meeting even got started. Shortly beforehand, an unsigned document was circulated to all the delegations apart from Japan describing the merits of Cadarache as well as many claimed shortcomings of Rokkasho, including the high cost of labour and electricity, risk of earthquakes and lack of infrastructure. The Japanese were furious. The US, following its recent return to the collaboration, was determined to get the site decision settled at the Washington meeting and so tried to calm frayed tempers.

The partners had been expected to have a sober debate and then come to an agreed position on where to build ITER. In the event, the meeting was a train wreck. First, the Canadians withdrew their site. Ontario Hydro and its partners had failed to win the support of the Canadian federal government and without that the site, and Canada's membership in the collaboration, were non-starters. With Cadarache and Rokkasho now head-to-head the partners lined up behind their favourites: Russia and China supported Cadarache; Korea and the US favoured Rokkasho. The negotiations became a slanging match. Europeans accused the US of only supporting Rokkasho because it couldn't stomach giving ITER to France, while the Americans charged Europe with blackmail after some French delegates said that if ITER went to Japan, France would pull out of the whole project. In the end, nothing was decided. The two sites were asked to provide more technical information to help resolve the issue. The champagne stayed corked and the teams headed home amid an air of mistrust and accusation.

For the next eighteen months, fusion scientists looked on in horror as their cherished project became ammunition for diplomatic warfare. Senior researchers were now excluded from meetings deciding the fate of ITER as government officials took over. After the conference-room sniping of the Washington meeting, salvos for and against the two sites continued in the media. US energy secretary Spencer Abraham told Japanese business leaders: 'From a technical standpoint you have offered the superior site.' The French prime minister Jean-Pierre Raffarin fired back: 'We have to have ITER, even if we do it ourselves...We won't let go of this.' High-ranking politicians toured the capitals of other ITER partners trying to win support. Both sides hinted that they would consider going ahead anyway with any partners that wished to join them. Japan and the EU even offered to shoulder larger and larger proportions of the total construction cost, in efforts to buy support.

Each side tried to exploit the weak points of the opposing site. Rokkasho's Achilles' heel was earthquake risk. Japan sits on the Pacific 'ring of fire,' an area around the ocean margin that is prone to quakes and volcanoes. Although Chinese officials didn't say anything publicly because they didn't want to inflame historic tensions between the two countries, they felt the seismic risk of Rokkasho was too great and hinted they would pull out if that site was chosen. Cadarache had the problem of being inland, so to get many of the reactor's huge components to the site, the project would have to widen roads, strengthen bridges and modify junctions along a winding 106-kilometre route. Japan argued that transporting such components by road was impractical if not impossible, but France cited the example of the Airbus A380 superjumbo. Although assembled in land-locked Toulouse, some of the plane's enormous parts, including whole wings and fuselage sections, are built elsewhere in Europe, shipped to southwest France and then trundled 240 kilometres through the countryside at night on purpose-built transporters.

Meanwhile, officials in Japan and France gathered more information on the suitability of the two sites in nine subject areas. A meeting was held in Vienna in mid March to debate the results. In France, licensing of the reactor would be covered by existing legislation and was already well underway; licensing in Japan required new legislation, a process which had not yet begun. The risk of a large-scale earthquake damaging to the reactor was considered twenty times more likely in Rokkasho than in Cadarache. Estimates put the cost of preparing the site at Cadarache at around one-eighth that at Rokkasho and the mild Mediterranean climate of Provence was certainly more appealing to researchers than the cold winters of northern Japan. In total, Cadarache was considered the better site in seven of the nine categories and in one they were judged equal. Its one failing was its inland position.

The comparison seemed to clinch it for Cadarache but the supporters of Rokkasho refused to concede and, because of their objections, the comparative document was never made public. The United States continued to insist it was supporting Rokkasho for technical reasons whereas in reality US officials backed Japan because they thought it would make a good, committed host, while it could not say the same of Europe. Some in the Bush administration didn't believe that the European Union could be treated in the same way as a sovereign state. Its twenty-five members had differing levels of commitments to ITER and the administration didn't think they could act with the unity of purpose that was needed to manage ITER's construction and pay for it.

Something had to give in the negotiations. The open hostilities were getting nowhere. Both sides realised that there was no way to resolve the issue while one side came out the 'winner' and the other the 'loser.' To save face, there had to be some prize that would go to the side that didn't get ITER. So began a series of bilateral negotiations between the EU and Japan, and the topic under discussion came to be known as the 'broader approach to fusion.' Fusion researchers had long recognised that to make progress towards a fusion power plant there were other things they needed to do apart from building ITER. They needed a particle accelerator facility to test the radiation hardness of materials that would be needed for such a plant, and supercomputers to simulate it. At that time, no one had any firm plans to build these facilities but now they were needed as bargaining chips. In order to cool the rhetoric between the two sides, the issue of who would have ITER and who the other facilities was put to one side – negotiators only referred to the 'host' and the 'non-host.' The hope was that if the facilities included in the broader approach became enticing enough, one side wouldn't mind having them rather than ITER.

The barrages of rhetoric calmed down as the potential host and non-host talked to each other. Mitsos was travelling to Tokyo twice a month during this period. Soon an appealing deal was worked out: the host would pay for almost 50% of ITER's total cost (with around 10% each from the other partners) and the non-host would get one or more expensive facilities from the 'broader approach' whose cost would be shared by the host and non-host. The problem was that both Japan and the EU still wanted to be the host. Something else was needed to tip the scales towards the non-host so that one of the two would be prepared to accept it.

One day in summer 2004, Rob Goldston, the director of the Princeton Plasma Physics Laboratory, was tidying his house. Japan's deputy science minister was coming to visit the lab and Goldston had invited him to dinner at his home on the evening before the visit. Once everything looked suitably welcoming,

Goldston sat on the stairs and tried to think of ways to break the deadlock over ITER's site because that evening's dinner provided him with a rare opportunity. Goldston knew that there is a strict protocol when dealing with Japanese politicians and some topics of conversation are off-limits. But there is also an unwritten rule that late in the evening, after a certain amount of alcohol has been consumed, it is acceptable to speak frankly and broach difficult subjects. Goldston had made a list of his ideas and when he was joined by his son Jake on the stairs he showed him the list. Jake, a university economics student, told his father, with the confidence of youth, that they were all useless, apart from number 4.

Idea number 4 was that as an added incentive for the non-host, the host – which is paying for 50% of the whole machine – would pay for some its components for ITER (for example, 10% of the total) to be built by firms in the non-host country. So the host still pays no more than 50%, but the non-host's industry gets more of the benefit. This, explained Jake, was the only one of Goldston's ideas in which the non-host got something unambiguously worth having in addition to the broader approach facilities. Goldston phoned officials at the Department of Energy and explained that he wanted to propose this idea to his Japanese visitor. He was told that it would be OK, so long as the idea was not attributed to the DoE.

The evening went according to plan. As soon as enough wine had been drunk, Goldston broached the subject of the ITER site and explained his idea. At the end of the evening the minister went away with a two-page memo spelling out the plan that Goldston had prepared earlier. When the visitor arrived at the lab the next morning, he immediately asked to use a fax machine. He wanted to send the memo to Tokyo before work ended that day. The ball was set in motion and would soon gather more momentum.

It took many more months for all the details to be worked out, but in May 2005 the *Yomiuri Shimbun*, a Japanese daily paper, quoted government sources as saying that Japan might be willing to give up its bid to host ITER if it won a lucrative role in construction. It took one final EU-Japan meeting the following week in Geneva to seal the deal. The two sides had resolved to have the issue settled before the 6th July start of the G8 economic summit at Gleneagles in Scotland. So at the beginning of July, just days before world leaders would gather in Scotland to discuss climate change and aid to Africa, and George Bush would collide on his bicycle with a British policeman, delegations from the ITER partners were welcomed to Moscow by Evgeniy Velikhov, the same person who twenty years earlier had persuaded Mikhail Gorbachev to propose ITER as a worldwide project 'for the benefit of all mankind.' Now he oversaw another turning-point in its progress: as everyone now expected, Cadarache was announced as the site for the reactor.

Also revealed was how much Europe had to pay to get it. The division of the ⇔5-billion cost followed the plan worked out between the EU and Japan including the extra 10% shifted from host to non-host outlined in Goldston's idea number 4. In addition, 20% of ITER headquarters staff would be Japanese and the EU would support Japan's proposal for a director general. As for the broader approach, Japan would get to choose a facility to build on its soil, up to a cost of ⇔800 million, with half paid by the EU.

After eighteen months of often rancorous negotiations, everyone seemed pleased with the outcome. Japanese industry could look forward to lucrative contracts paid for by Europe, while Europe could bask in the prestige of being the home of ITER. Although some European researchers worried about what they had taken on: the huge cost of ITER now threatened to starve all other fusion projects. Nevertheless, there was a palpable feeling of relief for everyone involved.

Just over a year later, France was able to chalk up one more

**A 1:50 scale model of the reactor at ITER headquarters in Cadarache.** (Courtesy of ITER Organisation)

minor victory over the United States when ministers from the seven partners (India joined early in 2006) came together to sign the international agreement that would make ITER an official collaboration. The ceremony was overseen not by President Bush in Washington but by President Jacques Chirac at the Elysée Palace in Paris. Earlier that year, Mitsos had stepped down from his job at the Commission and returned to Greece. His job was done. ITER was no longer a dream: it was a genuine collaboration of nations representing – with the addition of India – more than half the world's population. It now had a staff, a headquarters, a large patch of bare earth, and a plan. All they had to do now was build it.

CHAPTER 8

# If Not Now, When?

IN THE EARLY 1970S, WHEN MOST OF THE WORLD'S FUSION researchers were rushing to build tokamaks following the success of the Russian T-3 machine, some in the US thought it wasn't a good idea to turn their fusion programme into a one-horse race. In its twenty years of existence, the US programme had supported a range of different fusion machines – stellarators, pinches, mirror machines and other more exotic devices. But their number was decreasing. The stellarator had been largely abandoned with the arrival of the tokamak and others simply didn't perform well enough, suffering from instabilities, leaking plasma, or too low temperatures or densities. Something else was needed if the US programme wasn't going to become the tokamak show.

The strongest contender, though still some way behind tokamaks, was the mirror machine. These devices had been a mainstay at the Lawrence Livermore lab near San Francisco ever since it was founded and Richard Post was made head of its controlled fusion group. Post and his colleagues had built a number of small machines but instabilities were preventing them from confining a plasma for more than a fraction of a millisecond and plasma density remained low. But for reactor engineers, mirror machines have an elegant simplicity: straight field lines, plain circular magnets, predictable particle motions – a far cry from the geometrical contortions needed to make a tokamak or stellarator work. They confine plasma the same way that a stellarator does: with magnetic field lines aligned along the tube and particles pulled towards the lines so that they execute tight little spirals around them. The problems with a mirror start when the particles get to the end of the tubular vessel: how do you stop them escaping?

The simplest solution is to have an extra-powerful magnet coil around each end of the vessel. This squeezes the field lines into a tight bunch. When the spiralling particles encounter this more intense magnetic field they are repelled and head back the way they came, back along the tube. This works for the majority of particles but a lot still leak out, reducing the performance of the device. Researchers developed other more complicated designs for the end-magnet in the hope that they would produce a more leak-proof magnetic plug. These included coils in the shape carved out by the stitching on a baseball, and two interlocking versions of that shape, dubbed yin-yang coils in a nod to the Taoist symbol signifying shadow and light.

In 1973, researchers at Livermore were working with a mirror machine called 2XII and not getting very good results – the confinement was poor. They noticed, however, that containment improved when the plasma densities were higher, so they started looking for ways to inject more plasma and thereby boost the density. Colleagues at the nearby Berkeley Radiation Laboratory provided the solution: they were developing a system for producing neutral particle beams and to the 2XII team this seemed an ideal way to add more material to their plasma. It took two years to upgrade their machine for neutral beam injection but when they fired up their new 2XIIB – with an extra 'B' for 'beams' – in June 1975 they found that the containment was actually worse.

In desperation they tried a trick that had been suggested years before as a way of damping down instabilities: passing a stream of lukewarm plasma through the hot plasma in the vessel. The improvement was dramatic. By the following month they had doubled the ion temperature to 100 million °C – a record at the time – reached record plasma density and increased the confinement time tenfold. There was another side effect: the warm flow made the neutral beams very effective at heating the plasma, so much so that the researchers felt that they could abandon the heating method they had used previously – a rapid current pulse in the magnet coils to rapidly compress the plasma. Without the current pulse, 2XIIB was effectively a steady-state machine – the holy grail of reactor designers.

This really got the programme managers at the Energy Research and Development Administration (ERDA) interested. Here was something that could fill the role of a serious contender to the tokamak. It was little more than a year since the Middle East oil embargo had sent fuel prices through the roof and politicians were searching obsessively for anything that could become an alternative energy source – alternative to importing oil from the Arabian peninsula. Livermore hoped to jump on that gravy train and quickly drew up plans for a large-scale version of 2XIIB, to be called the Mirror Fusion Test Facility or MFTF, and the ERDA agreed to fund its construction.

While 2XIIB was undoubtedly a success, its end-magnets were still leaky and so many had doubts that MFTF would really make the grade as a power-generating reactor. Even using the most optimistic projections, a full-scale MFTF could only achieve a very modest gain – it would produce slightly more power than was used to keep it running. Hence there was considerable pressure on Livermore, during the design of MFTF, to come up with some technique to stop the leaks and improve the gain.

A solution was found in 1976, simultaneously with researchers in the Soviet Union, with a system that came to be known as a tandem mirror. In such a machine, the single magnets at each end are replaced by a pair of magnets separated by a short straight section. The two magnets and the plasma confined between them form a 'mini-mirror' system and this proved to be a more effective plug than a single mirror alone. To test the idea, Livermore persuaded ERDA to fund the construction of another experiment, smaller than MFTF, called the Tandem Mirror Experiment. They built TMX and sure enough they showed that the double-magnet end plugs reduced leakage. The only trouble was that now the Livermore researchers had to redesign MFTF. They chose to shorten the existing MFTF and make it into one of the end plugs, so they now needed another whole MFTF as the second end plug and a new central section. The new design, dubbed MFTF-B, was significantly larger and more expensive than the original one but, having got this far, ERDA agreed to press ahead.

While tandem mirrors reduced leakage, there was still room for improvement. In 1980, researchers at Livermore came up with another idea: put a third magnet at each end of the machine and this would produce a double plasma plug at both ends to further block escaping plasma. To test the idea they upgraded the TMX machine with the extra magnets and it did produce a more effective plug. That in turn led to another redesign of MFTF-B to add extra magnets.

By now MFTF-B had turned into a monster of a machine. The plasma tube and all the magnets at each end were enclosed in a stainless steel vacuum vessel that was 10m in diameter and 54m long. You could easily drive a double-decker bus down the middle of it. When it started operating it would need a staff of 150 to tend to it and was expected to consume $1 million of electricity every month. The elaborate end plugs were far from the simple mechanisms that had attracted many people to mirror machines in the first place. Some researchers joked that, like a tree's rings, you could tell how old a mirror machine was by how many magnets had been added to the ends.

In all it took nine years to build MFTF-B at a cost of $372 million. On 21st February, 1986 staff and guests gathered for the official dedication ceremony. The Secretary of Energy, John Herrington, had travelled over from Washington along with other DoE staff and he commended the Livermore team for their work. But it wasn't the joyous occasion everyone had been expecting.

The political climate in the mid 1980s was very different from a decade earlier. Ronald Reagan had come into the White House in 1981 and had aggressively cut public spending. The high oil prices and frantic search for alternative energy sources of the 1970s were now just a memory. To the Reagan-era DoE, funding a second type of fusion reactor just to provide competition for tokamaks was an expensive luxury. So the day after congratulating Livermore on its achievement, DoE shut the doors on MFTF-B without ever having turned it on. A few years later it was dismantled for scrap and to this day the scientists, engineers and technicians who spent years working on the machine do not know if it would have worked.

What is the moral of this story? Fusion energy isn't inevitable. No fusion machine, no matter how much has been spent on it, is safe from the budget axe. The giant machines of today – the newly completed National Ignition Facility and the partially build ITER – could suffer a similar fate to MFTF-B. The search for fusion energy is expensive and it will only continue if politicians and the public want it and need it.

Twenty-three years later on 31st March, 2009 Livermore held another dedication ceremony, this time for the National Ignition Facility. NIF wasn't about to be shut down but it was under enormous pressure to perform. NIF's funders at the National Nuclear Security Administration (NNSA), a part of the Department of Energy, wanted payback for the huge cost of the machine, and wanted it soon. A few years earlier, while NIF was still being constructed, NNSA officials and senior researchers drew up a plan to get to ignition on NIF as quickly as possible, so as to provide a springboard for all the things they planned to do with the facility: weapons research, basic science and, of course, fusion energy.

Called the National Ignition Campaign (NIC), it began in 2006 and included designing the targets for NIF shots and simulating what was likely to happen to those targets. Other labs were also involved in the NIC, including Rochester University's Laboratory of Laser Energetics and its Omega laser as well as the Z Machine at Sandia National Laboratory which studies inertial confinement fusion with very high current pulses rather than lasers. These facilities were able to try out at lower energy some of the things that would eventually be done at NIF.

By the time of NIF's inauguration the NIC was three years old and researchers were confident that ignition was within their reach. There was a lot of calibrating and commissioning to do on NIF so it would be well into 2010 before researchers could do shots on targets filled with deuterium-tritium fuel that would be capable of ignition, but the NIF team said they may well reach their goal before the end of that year.

NIF is a truly astounding machine. Its size alone takes your breath away: the building that contains it is the size of a football stadium and ten stories high. Inside is a laser so big and so powerful it would make a James Bond villain weep. Among all the brushed metal, white-painted steel superstructure and tidily bundled cables there is the hum of quiet efficiency. The place has a feeling of a huge power, ready to be unleashed.

The heart of the machine is a small unassuming optical fibre laser that produces an unremarkable infrared beam with an energy measured in billionths of a joule. This beam is split into forty-eight smaller beams and each is passed through a separate preamplifier in the shape of a rod of neodymium-doped glass. Just before the beam pulse arrives, the preamplifiers are pumped full of energy by xenon flashlamps and this energy is then dumped into the beam as it passes through. After four passes through the preamplifiers the energy of the forty-eight beams has been boosted ten billion times to around 6 joules. The beams are then each split into four beamlets – giving a total of 192 – and passed through to NIF's main amplifiers. These are what takes up most of the space in the facility's cavernous hall, being made up of 3,072 slabs of neodymium glass (each nearly a metre long and weighing 42 kilograms) pumped by a total of 7,680 flashlamps. After a few passes through the amplifiers the beams, which now have a total energy of 6 megajoules, head towards the switchyard.

The switchyard is a structure of steel beams which supports ducts and turning mirrors to direct the beams all around the spherical target chamber so that they all approach from different directions. The chamber itself is 10m in diameter and made from 10cm-thick aluminium with an extra 30cm jacket of concrete on the outside to absorb neutrons from the fusion reactions. This protective sphere is punctured by dozens of holes: square ones for the laser beams and round ones to act as viewing ports for the numerous diagnostic instruments that will study the fusion reactions. Before the beams enter the chamber they must pass through one final but crucial set of optics. These are the KDP crystals which convert the infrared light produced by the Nd:glass lasers, with a wavelength of 1,053 nanometres, first to green light (527 nm) and then to ultraviolet light (351 nm) because this shorter wavelength is more efficient at imploding fusion targets.

Finally the beams, which for most of their journey have filled ducts 40cm across, are focused down to a point in the middle of the target chamber where they must pass through a pair of holes in the hohlraum each 3mm across. For all its size and brute force,

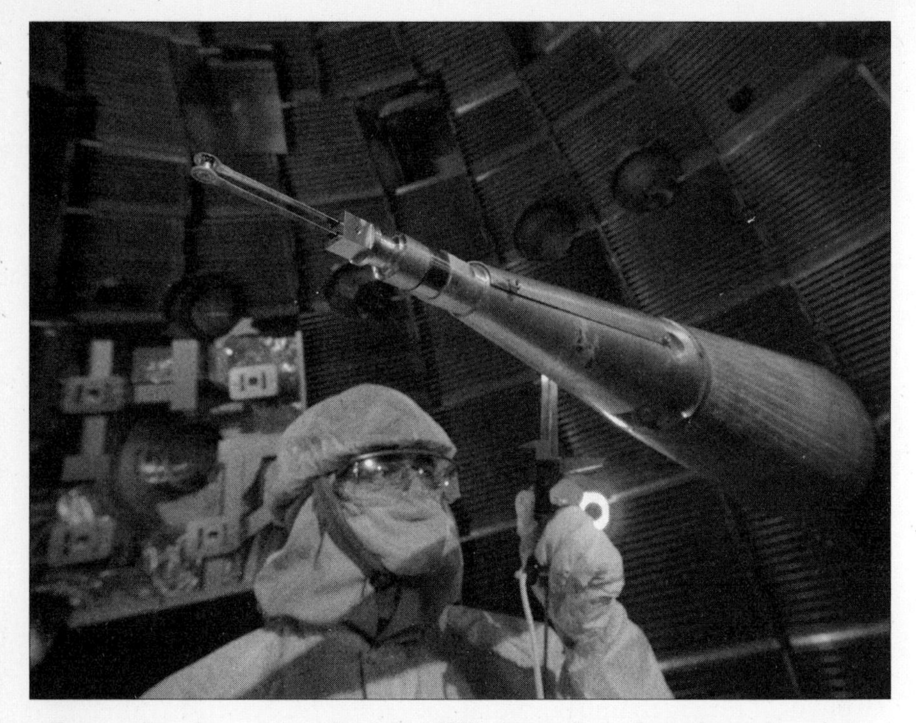

**Preparing for a shot: Inside NIF's reaction chamber showing part of the positioner arm and, at its tip, a hohlraúm.**
(Courtesy of Lawrence Livermore National Laboratory)

the laser's end result must be needle-fine and extremely precise. The hohlraum is held in the dead centre of the target chamber by a 7m-long mechanical arm. This positioner must hold the target absolutely steady in exactly the right spot with an accuracy of less than the thickness of a piece of paper. The arm also contains a cooling system to chill the target down to -255°C so that the deuterium-tritium fuel freezes onto the inside wall of the capsule.

An NIF shot goes like this: the original fibre laser creates a short laser pulse, around 20 billionths of a second long, which then travels through the preamplifiers, amplifiers and final optics before it enters the target chamber as 192 beams of ultraviolet light with a total energy of 1.8

megajoules. This is roughly equivalent to the kinetic energy of a 2-tonne truck travelling at 160 kilometres/hour (100 miles/hour) but because the pulse is only a few billionths of a second long its power is huge, roughly 500 trillion watts which is 1,000 times the power consumption of the entire United States at any particular moment. With that sort of power converging on the hohlraum, things start happening very fast. The 192 beams are directed into the hohlraum through holes in the top and bottom and onto the inside walls. The walls are instantly heated to such a high temperature that they emit a pulse of x-rays. The hohlraum's interior suddenly becomes a superhot oven with a temperature of, say, 4 million °C and x-rays flying all over the place. The plastic wall of the capsule starts to vaporise and flies off at high speed. This ejection of material acts like a rocket, driving the rest of the plastic and the fusion fuel inwards towards the centre of the capsule.

If the NIF scientists have got everything right, then this inward drive will be completely symmetrical and the deuterium-tritium fuel will be crushed into a tiny blob around 30 thousandths of a millimetre across and with a density 100 times that of lead. The blob's core temperature will be more than 100 million °C but even this isn't quite enough to start fusion. The laser pulse has a final trick up its sleeve to provide the spark. If the timing is right then a converging spherical shockwave from the original laser pulse should arrive at the blob's central hot spot just as it reaches maximum compression. This shock gives the hot spot a final kick to start nuclei fusing. Once the reactions start, the high-energy alpha-particles produced by each fusion heat up the slightly cooler fuel around the hot spot. That leads to more fusions, more alpha particles, the reaction gains its own momentum and – BOOM! – fusion history is made. A faultless shot might produce 18 megajoules of energy, ten times that of the incoming laser beams.

That sequence of events was the goal when Livermore researchers began their NIC experiments in 2010. The first unknown was whether the laser was up to the job. Critics of NIF had warned that laser technology wasn't ready for such a high-energy machine. They foretold that the amount of power moving through the optics would cause them to overheat and crack; that specks of dust on glass surfaces would heat up and damage them; and that flashlamps would continually blow out and need to be replaced. During NIF's construction there were problems with exploding capacitor banks and flashlamps, and the whole system for preparing and handling the optical glass had to be redesigned to keep it as clean as a semiconductor production plant. NIF's designers did their work well. When they finally turned it on and ramped up the power over the first couple of years, the laser didn't tear itself to pieces. The occasional lamp did blow and some optical surfaces got damaged, but NIF staff had worked out ways to either repair surfaces or to block out a damaged section so that the laser could keep running.

The Livermore researchers knew from earlier machines that getting the laser to work was just the start: numerous hurdles still lay ahead. The first of these was the chaotic environment inside the hohlraum once the laser pulse starts. While the high-energy beams are heating the inside walls they kick up lots of gold atoms that form a plasma inside the hohlraum. If it's not carefully controlled this plasma can cause havoc, sapping the energy of the incoming beams, diverting them from their desired paths and even reflecting some of the beam back out of the hole in the hohlraum. Such interactions had limited the achievements of earlier laser fusion machines and the NIF team studied the problem carefully and carried out extensive simulations. In the early experiments of the NIC the team mostly managed to keep these plasma interactions under control, largely by avoiding situations that were known to aggravate them.

Another potentially difficult area was the implosion of the capsule. The implosion is an inherently unstable situation because it involves a dense material – the plastic shell – pushing on a less dense one – the fusion fuel, and Rayleigh-Taylor instabilities can lead to fuel trying to burst out of its confinement. Researchers' first weapon against this is symmetry, hence the careful placement of beams around the hohlraum interior to ensure that the x-rays heat the capsule evenly. Their second weapon is speed: if they can make the implosion sufficiently fast, the plastic and fuel won't have time to bulge out of shape.

The experimenters use a measure called the experimental ignition threshold factor (ITFX) to chart their progress. The ITFX is defined so that an ignited plasma has an ITFX of 1. For the first year of the NIC, the value of ITFX demonstrated the advances they made. When ignition experiments started the shots achieved an ITFX value of 0.001. A year later it had reached 0.1 – a hundred-fold increase – but there it stalled. The second year of the NIC was plagued by phenomena that the NIF team was unable to explain. Although the target chamber was bristling with diagnostic instruments – nearly sixty in total – to probe what was going on inside, measuring x-rays, neutrons, and even taking time-lapse movies of the implosions, the researchers could not figure out why the capsules were not behaving as the computer simulations said they would. For reasons unknown, a significant portion of the laser beam's energy was getting diverted from its intended purpose of driving the implosion of the capsule. The capsule shell was also being preheated before the implosion started – perhaps by the stray laser energy – which made it less dense and less efficient at compressing the fuel. The implosion velocity was also too slow. At a fusion conference in September 2011, the DoE's Under Secretary for Science, Steven Koonin, who oversaw NIF, said that 'ignition is proving more elusive than hoped.' He also said that 'some science discovery may be required,' which is a polite way of saying 'we don't know what's going on.'

Koonin set up a panel of independent fusion experts to give

him regular reports on NIF's progress. The panel was critical of the schedule-driven approach of the NIC, which specified what shots had to be carried out and when and, if something unexpected arose, didn't allow any time to explore what was going wrong. In a report from mid 2012 the panel pointed out that Livermore's simulations of NIF predicted that the shots they were then carrying out should be achieving ignition, but the measured ITFX values showed they were still a long way off. What sort of a guide to progress were the simulations if their predictions were so wide of the mark? As before in laser fusion, it was simulations that led researchers to have inflated expectations.

It had been stipulated when the NIC was devised that if ignition is not achieve by the end of two years of experiments at NIF then the NNSA had sixty days to report to Congress on why it had failed, what could be done to salvage the situation, and what impact this will have on stockpile stewardship. That deadline passed on 30th September, 2012 and on 7th December the NNSA submitted its report to Congress. The report admitted that Livermore researchers did not know why the implosions were not behaving as predicted and even conceded that it was too early to say whether or not ignition could ever be achieved with NIF. The NNSA asked for NIF's funding – running at roughly $450 million per year – to be continued for a further three years so that researchers could investigate why there was a divergence between simulations and measured performance. Significantly, the report also called for parallel research to be carried out on other approaches to ignition as a backup in case NIF failed. These alternatives included pulsed-power fusion at Sandia's Z Machine, direct-drive laser fusion using Rochester's Omega and even direct drive on NIF.

At the time of writing, it was not known how Congress would react to this proposal although President Barack Obama's proposed budget for 2014 suggests cutting NIF funding by 20%. Also, some members of Congress have campaigned for years for NIF's closure and this admission of weakness could only help their cause. Whatever happens, progress towards ignition looks set to slow because, while the NIC used around 80% of the shots on NIF, from the beginning of 2013 weapons scientists would be getting a bigger share, more than 50%. Many laser fusion experts still believed that NIF can get to ignition, but the question is: when?

Meanwhile, in France, the ITER project was just starting to get moving. Following the ceremony in Paris to sign the international agreement in November 2006 it took almost a year for each partner to ratify the treaty and only then could they officially create the ITER Organisation. But that didn't hold up excavation of the site. Machinery began clearing trees from land near Saint Paul lez Durance in January 2007. Some rare plants and animals were moved elsewhere; remains of an eighteenth-century glass factory and some fifth-century tombs were preserved. Heavy earth-moving machines arrived in March 2008 and set about carving away at the side of a hill to lower the ground level then shifting the earth downhill to build it up into a level surface. Altogether the diggers shifted 2.5 million cubic metres of material to create a platform of 42 hectares, the area of sixty football pitches. And then everything came to a halt as the new team in the temporary office buildings nearby struggled to get to grips with the machine they had to build.

It hadn't taken long after the final decision to build ITER at Cadarache for the project partners to start filling senior management positions. As expected, Japan's choice for director general, Kaname Ikeda, the country's ambassador to Croatia, was approved. Ikeda had held numerous government jobs relating to research and high-tech industry and had a degree in nuclear engineering – a suitably senior person to head such an international organisation, but not a fusion scientist. His deputy was Norbert Holtkamp, a German physicist who had a track record of managing big projects, having just finished building the particle accelerator for the Spallation Neutron Source at Oak Ridge National Laboratory in Tennessee – but again, not a fusion scientist.

These two couldn't have been more different. Ikeda was every inch the Japanese diplomat: polite, deferential, immaculately presented. Holtkamp, by contrast, was laid-back, affable and liked to do things his own way. Even a few years working in the United States hadn't squeezed him into the corporate mould – at ITER he managed to persuade the office manager to exempt him from the usual no-smoking rules so that he could puff cigars in his office. Below them were seven deputy directors, one from each project partner, and the recruitment continued in a similar vein, trying to keep a similar number of staff from each partner. It was a management structure designed by a committee of international bureaucrats and it would prove to be a millstone around the young organisation's neck.

This new management team, whose leaders were new to fusion, included many researchers who had not been involved in drawing up the design for ITER, so the first thing they had to do was thoroughly familiarise themselves with the machine. They also had to check and recheck every detail of the design to ensure it was ready to be used as the blueprint for industrial contracts to build the various components of the reactor. Then there was the fact that the design was, by then, half a dozen years old and fusion science had moved on: now was the chance to incorporate the latest thinking into the design. So the new team appealed to the worldwide fusion community to come to their aid. They asked researchers to fill in 'issue cards' describing any aspect of the design that worried them or possible improvements that could be applied.

Fusion scientists weren't shy in coming forward and by early 2007 the ITER team had received around 500 cards. They drafted in outside experts to help and set up eight expert panels to sift through all the concerns and suggestions. Many proved to be impractical and could be discounted, others required just minor modifications to the design, but a few required large – and expensive – changes. By the end of 2007 they had whittled the list down to around a dozen major issues and work still remained to figure out how these could be incorporated into the design without inflating the cost.

One of the most contentious concerned a new method for controlling edge-localised modes (ELMs), the instability that causes eruptions at the edge of the plasma that can damage the vessel wall or divertor – a side-effect of the superior confinement of H-mode. The solution already chosen for ITER was to fire pellets of frozen deuterium into the plasma at regular intervals. These cause minor ELMs, letting some energy leak out and so preventing larger, more damaging ones. But researchers working on the DIII-D tokamak at General Atomics in San Diego had come up with a simpler solution: applying an additional magnetic field to the plasma surface roughens it up, which also allows energy to leak out in a controllable way. The problem was that this additional field required a new set of magnet coils to be built on or close to the vessel wall, a potentially costly change at this late stage.

The ultimate goal of these early years was to produce a document called the project baseline. Thousands of pages long, the baseline is a complete description of the project, including its design, schedule and cost. The project partners were not going to sign off on the start of construction until they had seen and approved the baseline document. So the levelled site sat quiet and empty while the ITER team continued to wrestle with a paper version of the project.

In June 2008 the team presented the results of the design review to the ITER council, which is made up of two representatives from each partner. Despite their extensive whittling down of the many suggestions from researchers there were still numerous refinements and modifications to components including the main magnets and the heating systems, plus there were the additional magnet coils to control ELMs. These changes, the team estimated, would add around ⇔1.5 billion to the cost. Nor did the original estimate of ⇔5 billion for construction look that secure. The more the ITER team looked into the details of the design, the more they found that the designers of 2001 may have been overly optimistic and that the final cost could be as much as twice the original estimate.

It's not uncommon for major scientific projects such as ITER to go over budget, but this ballooning price tag would be an especially hard sell because, since the founding of ITER in 2006, the worldwide financial crisis had swept through the economies of all the project partners. The delegations would not relish going back to their governments and asking for more money when 'austerity' was the new black. So the council sent the ITER team back to work with the instruction to nail down the cost completely so that there would be no more surprises. Nor did the council trust them to get on with it unsupervised. It appointed an independent panel, led by veteran Culham researcher Frank Briscoe, to investigate the project cost and how it was estimated. It also set up a second panel to study the organisation's management structure.

A year later the team asked the council for permission to build ITER in stages. First they would fire up the machine with just a vacuum vessel, magnets to contain the plasma, and the cryogenic system needed to cool the superconductors in the magnets. The idea behind this was to make the whole system simpler so that operators could get the hang of running the machine without all the added complication of diagnostic instruments, particle and microwave heating systems, a neutron-absorbing blanket on the walls and a divertor, which would be added later. The council agreed and also approved a slip in the schedule: first plasma in 2018 not 2016, and first D-T plasma in 2026, nearly two years later than originally planned. The ITER team were still working on the project baseline, including that all-important cost estimate, but the council asked to see it at its next meeting in November 2009.

But the autumn meeting ended up being all about schedule. The European Union was concerned that finishing construction by 2018 was still too soon. Rushing the process could lead to mistakes that would be impossible to correct later. So ITER's designers were sent back to the drawing board to do more work on the schedule. Other members of the collaboration were getting frustrated by the delays. They wanted to push ahead as quickly as possible but Europe, as the project's biggest contributor at 45%, had the most to lose and so could throw its weight around. In the spring of 2010 a new completion date was agreed: November 2019. But still there was no baseline and the reactor's home remained an idle building site.

Later that spring the true underestimation of ITER's cost became apparent. The European Union was going to have to pay ⇔7.2 billion for its 45% share of the cost. That put the total bill in the region of ⇔16 billion. While this colossal figure would impose a severe financial strain on fusion funding for all the ITER partners, for the EU it caused a near meltdown. The problem was this: EU funding is agreed by member states in seven-year budgets; the current budget cycle ran until the end of 2013; the budget line for fusion in the years 2012 and 2013 contained ⇔700 million; but the new inflated ITER cost required ⇔2.1 billion from the EU in those years, so Europe had to find another ⇔1.4 billion from somewhere.

EU managers considered a number of options to fill the gap, including getting a loan from the European Investment Bank, an EU institution that lends to European development projects. But this was rejected because there was no identifiable income stream to repay the loan. They considered raiding the budgets of EU research programmes but feared a backlash from scientists across the continent. In the end, they appealed directly to EU member governments for an extra payment to get them out of a hole. In June member states declined to bail out the project, essentially saying this was the EU's problem and it would have to find its own solution. The June ITER council meeting came and went with still no decision on the baseline.

With much persuasion and institutional arm-twisting, EU managers finally managed to cobble together the necessary funds from within the EU budget. Some ⇔400 million was taken from other research programmes and the rest from other sources, in particular unused farming subsidies. The way was now clear for ITER to move forward.

On 28th July, 2010, the ITER council met in extraordinary session at Cadarache. Chairing the meeting was none other than Evgeniy Velikhov, again on hand to guide ITER through one of its major turning-points. With huge relief the national delegates approved the baseline, allowing ITER to move into its construction phase. But that was not their only item of business. They also bade farewell to director general Kaname Ikeda, who had asked to stand down once the baseline was approved. In his place, the council appointed Osamu Motojima, former director of Japan's National Institute for Fusion Science. Motojima knew how to build large fusion facilities, having led the construction of Japan's Large Helical Device, a type of stellarator. Ikeda's was not the only departure following the baseline debacle. Norbert Holtkamp also stood down and his position of principal deputy director general was dispensed with.

Motojima's mandate from the council was to keep costs down, keep to schedule and simplify ITER management. The latter he set about by sweeping away the previous seven-department structure and replacing it with a more streamlined three. The first department, responsible for safety, quality and security, was

**Evgeniy Velikhov celebrates the end of his term as ITER council chair in November 2011.** (Courtesy of ITER Organisation)

headed by Spaniard Carlos Alejaldre, who had filled the same role under the old structure. To run the key ITER Project Department, responsible for construction, Motojima appointed Remmelt Haange of Germany's Max Planck Institute for Plasma Physics. Haange, like Motojima, was a seasoned reactor builder, having been involved in the construction of JET and as the technical director of Germany's Wendelstein 7-X stellarator project. Finally, Richard Hawryluk, deputy director of the Princeton Plasma Physics Laboratory, was picked to head the new administration department. With these veteran fusion researchers at the helm, the project got down to the serious business of building the world's biggest tokamak. Soon trucks and earth-moving machines were crawling over the site like worker ants.

Research into fusion energy is now well into its seventh decade. Thousands of men and women have worked on the problem. Billions have been spent. So it seems reasonable to ask, will it ever work? Will this amazing technology, which promises so much but is so hard to master, ever produce power plants that can efficiently and cheaply power our cities? Today's front-rank machines, such as NIF and ITER, seem so thoroughly simulated and engineered, and the previous generation of machines got so close to breakeven, that surely the long-sought goal can't be far away? Let's first consider the case of inertial confinement fusion and NIF.

At the time of writing it is still anyone's guess whether NIF will ever be made to work. Many believe that it is just a matter of twiddling all the knobs until the right combination of parameters is found and suddenly everything will gel. But Livermore's choices of neodymium glass lasers and indirect drive targets have always been controversial and critics say the whole field needs a complete change of direction, such as to krypton-fluoride gas lasers – which are naturally short wavelength – and simple and cheap direct-drive targets.

NIF cannot escape from the fact that its primary goal isn't fusion energy but simulating nuclear explosions to help maintain the weapons stockpile. Nevertheless, when the machine was inaugurated in 2009 the press coverage focused almost exclusively on fusion energy. That was no accident. During the preceding years NIF's managers felt which way the political wind was blowing: while maintaining the nuclear stockpile was still important, so was climate change and energy independence. If they were to maintain support for NIF from the public and Congress they had to broaden its appeal. Hence the emphasis on energy, not nukes.

NIF director Ed Moses and his team expected that when they achieved ignition it would spark a surge of interest in fusion energy and – hopefully – new money. They wanted to be ready to ride that wave of enthusiasm so, in traditional fashion, they started to plan for the reactor that would come next, one designed for energy production, not science or stockpile stewardship. Taking the achievement of ignition – which they expected soon – as their starting point, they sought to establish how fast and how cheaply they could build a prototype laser fusion power plant. They adopted a deliberately low-risk approach, sticking as closely as possible to NIF's design in order to cut down on development time. All components had to be commercially available now or in the near future. They consulted with electricity utility companies about what sort of reactor they would like – something fusion researchers had never really done before. They called this dream machine LIFE, for laser inertial fusion energy.

The first thing to tackle was the laser. NIF's laser, though a wonder, is totally unsuitable for an inertial fusion power plant. It's a single monolithic device, prone to optical damage, and could only be fired a few times a day. The NIF team didn't want to abandon neodymium glass lasers altogether – it was the technology that they knew and understood. But they could get rid of the temperamental and power-hungry xenon flashlamps that pump the laser glass full of energy. The ideal alternative would be solid-state light emitting diodes, similar to those used in LED TV screens and the latest generation of low-energy light bulbs. They are more efficient than flashlamps, power up more quickly and are less prone to damage. Electronics companies can make suitable diodes today but they are so expensive that they would make a laser power plant uneconomic. However the NIF researchers calculated that, like most electronic components, their price will go down rapidly and, by the time LIFE needs them, they will be affordable.

It also wouldn't do to have LIFE relying on a single laser to drive the whole power plant. If any tiny thing went wrong the entire plant would have to be shut down for repairs. So instead of a single laser split into 192 beams, LIFE would have twice as many beams (384) with each produced by its own laser. The plan was for the lasers to be produced in a factory as self-contained units – essentially a box big enough to keep a torpedo in. The operators of the plant wouldn't need to know anything about lasers; the units would be delivered by truck and the operators would just slot them into place and turn them on. The plant would have spares on site and if one laser failed it could be pulled out and be replaced without stopping energy production.

Livermore researchers also had a novel solution to one of the big questions of fusion reactor design: neutron damage. Nuclear engineers are working hard to find new structural materials for fusion reactors that can withstand a constant barrage of high-energy neutrons for years on end. But the Livermore team didn't want to have to wait for new materials to be developed and tested. They opted for a simpler solution for LIFE: make the reaction chamber replaceable. In their design, the only thing that physically connects to the reaction chamber is the pipework for cooling fluid. After a couple of years this can be disconnected and the entire reaction chamber wheeled out on rails to an adjacent building, then a fresh chamber can be wheeled in. The old chamber would need a few months to 'cool off' so that its radioactivity drops to a safer level, then be dismantled and buried in shallow pits.

It was a bold plan and, because of its policy of relying on known technology and off-the-shelf components, the team calculated that, once ignition on NIF is achieved, they could build a prototype LIFE power plant in just twelve years.

Livermore wasn't the only organisation thinking ahead to what will happen after ignition is achieved. The US Department of Energy and in particular its science chief Steven Koonin realised that when that breakthrough came the White House, Congress, other organisations and the public would start asking questions, such as what has the US been doing in inertial fusion in recent years? And what is it going to do now to progress from scientific breakthrough to commercial power plant? The answer to the first question was: not very much. During the construction of NIF and afterwards, other research on inertial confinement fusion was starved of funding. The Rochester University lab got some money for its supporting role to NIF but research at the national laboratories and elsewhere was minimal. However that didn't mean that those few researchers working in the field didn't have ideas about what to do next.

Koonin was very familiar with NIF, having been drafted onto various panels over the years to assess it and other fusion projects. What Koonin needed now was a broad survey of the state of the whole field of inertial fusion research, so again the National Academy of Science was called on to investigate. The NAS assembled a panel of experts from universities, national labs and industry. Over the course of a year they visited many of the main facilities involved in inertial confinement fusion research and heard dozens of presentations. Their first port of call outside Washington was to Livermore where NIF researchers explained their plans for the LIFE power plant.

Next they visited the Sandia National Laboratory in Albuquerque, New Mexico. Researchers there had been working on inertial confinement fusion using, not lasers, but extremely powerful current pulses to crush a target magnetically. Their technique uses the pinch effect, the same phenomenon that Peter Thonemann stumbled upon in the 1940s and caused him to travel from Australia to Oxford to start building fusion reactors. The pinch effect causes a flowing electric current to be squeezed by its own magnetic field inwards towards its middle. Thonemann's devices, along with all tokamaks, use the pinch effect to squeeze a flowing plasma, compressing and heating it. The researchers at Sandia use the pinch in a different way. They confine fusion fuel in a cylindrical metal can and then pass a huge current down the outer walls of the can. The pinch effect squeezes the walls of the can in towards the centre and so crushes the can. If the current pulse is big enough and fast enough then the crushed can compresses and heats the fuel inside enough to spark fusion.

To do this requires an enormous pulse of electric current and so Sandia researchers use the Z Machine which can store charge in huge banks of capacitors and then release it quickly. The 37m wide machine can create current pulses of 27 million amps lasting a ten-millionth of a second. In 2013 the Sandia team will start trying to achieve fusion using the Z Machine and simulations suggest they might be able to reach break-even. But to really put the idea to the test and produce genuine energy gain they reckon that they need a new machine – Z-IFE – able to produce current pulses up to 70 million amps.

As a potential power source, the Z Machine has the drawback that it takes longer to charge up than a laser and the metal can targets are bigger and more cumbersome than the fuel capsules of laser fusion. So the Sandia scheme would work at a slower repetition rate – 1 shot every 10 seconds. To make that rate economic, each explosion would have to be bigger to generate more power, so this scheme has the added challenge of developing a reaction chamber that can withstand a much bigger blast and be ready for the next one every 10 seconds. Nevertheless, the team believes that their sledgehammer approach of big pulses, big bangs and slow repetition is much simpler to implement than the high speed and pinpoint accuracy required for laser fusion energy.

The NAS panel's next port of call was Rochester and the Laboratory of Laser Energetics. Here the main topic of discussion was other ways to achieve fusion with lasers that avoid some of the drawbacks of NIF's indirect drive using neodymium-glass lasers. Researchers at Rochester and the Naval Research Laboratory in Washington, DC, argue that direct drive would be a better approach for laser fusion energy generation. Shining the laser directly onto the fuel capsule avoids the energy lost in the hohlraum, so a less powerful laser would be needed. It would be simpler too, since instead of having to construct a target with a fuel capsule carefully positioned inside a gold or uranium can for each shot, only the fuel capsule would be needed. Since future laser fusion power plants are expected to have to perform around 10 shots per second they will consume slightly less than a million targets a day, so simplicity – and just as importantly, low cost – will be a key factor.

Livermore researchers abandoned direct drive because you need laser beams of very high quality to make it work; any imperfections in the beam and the capsule will not implode symmetrically. But Rochester and the Naval Research Lab stuck at it and developed ways to smooth over beam imperfections. They have tested these techniques using Rochester's Omega laser but it doesn't have the power of NIF so they have not been able to test direct drive to ignition energies.

The Naval Research Lab team have also made another innovation. They have developed a laser that emits light that is already ultraviolet so it doesn't need to be stepped down to shorter wavelengths like Omega and NIF, which avoids the energy loss of conversion using KDP crystals. Instead of using neodymium-doped glass as the medium to amplify light, their laser uses krypton-fluoride gas which is pumped full of energy using electron beams.

The Naval researchers have built demonstration models that have a high repetition rate but low power, and ones with high power but are only capable of single shots. They haven't yet won funding to develop a high-power, high-repetition version ready for fusion experiments.

The NAS panel heard about other approaches as well, such as imploding targets using beams of heavy ions. Accelerating ions is a much more energy-efficient process than making a laser beam and there is no problem with creating a high repetition rate. Focusing also uses robust electromagnets rather than delicate lenses which can get damaged by blasts or powerful beams. But creating beams with the right energy and sufficiently high intensity to implode a target is still a challenge. Researchers at the Lawrence Berkeley National Laboratory near San Francisco have built an accelerator to investigate those challenges but the project is desperately short of funds.

Then there is another laser-based technique called fast ignition. This separates out the two functions of the laser pulse – compressing the fuel and heating it to fusion temperature – and uses a different laser for each. In a conventional laser fusion facility these two jobs are achieved by carefully crafting the shape of the laser pulse during its 20-nanosecond length: the first part of the pulse steadily applies pressure to implode the capsule and compress the fuel, then an intense burst at the end sends a shockwave through the fuel which converges on the central hotspot, heating it to the tens of millions of °C required to ignite. Getting that pulse shape right is complicated and requires a very high-energy laser. By separating the two functions a fast ignition reactor can make do with a much lower-energy driver laser because it only has to do the compression part. Once the implosion has halted and the fuel has reached maximum density, a single beam with a very short but very high intensity pulse is fired at the fuel and this heats some of it to a temperature high enough to spark ignition.

Researchers at Osaka University in Japan have pioneered the fast ignition approach and have been joined recently by Rochester's Omega laser which was upgraded with a second laser for fast ignition experiments. Although these efforts are making progress towards understanding how fast ignition works, neither is powerful enough to achieve full ignition. Researchers in Europe are keen to join this hunt and have drawn up detailed plans for a large fast ignition facility that would demonstrate its potential for power generation. Known as the High Power Laser Energy Research facility (HiPER), it would have a 200-kilojoule driver laser – one-tenth the energy of NIF's – and a 70-kilojoule heater laser. Calculations suggest HiPER should be able to achieve a much higher energy gain than NIF but its designers are waiting to finalise the design. They want to wait for the achievement of ignition at NIF in case it provides any useful lessons.

The problem for all these possible alternatives – with the possible exception of Rochester's Omega laser – is that they have been starved of funding. For the past two decades while the Department of Energy has been pouring money into NIF other approaches have been neglected. The fear for the proponents of these alternative approaches was that history was about to repeat itself: Livermore's design for the LIFE reactor is very thorough and very persuasive; would it persuade the NAS panel that the bulk of any future ICF funding should be channelled straight to Livermore?

But the panel was not seduced. Its mandate had been to come up with a future research programme *on the assumption that ignition at NIF had been achieved*. Ignition's stubborn refusal to cooperate at NIF has knocked some of the lustre off the plans for LIFE – it no longer seemed an obvious shoo-in for the next big inertial fusion project. The panel's report, released in February 2013, says that many of the technologies involved in inertial fusion are at an early stage of technological maturity and that it is too early to pick which horse to back. It suggested a broad programme of research that would provide the information needed to narrow the field in the future. This was good news for the alternative approaches, but with a faltering US economy forcing cutbacks to many areas of research funding and with Livermore researchers seemingly losing their way on the road to ignition, prospects for a generous research programme do not look good.

With the construction of ITER going at full throttle and with NIF working towards ignition, is fusion closing in on the big breakthrough that researchers have been dreaming of for more than six decades? Many thousands of those researchers would like to believe so but there have been false dawns before: Ronald Richter's Argentine fusion reactor that never was; ZETA and all the enthusiasm generated by the 1958 Geneva conference; the astonishing temperatures achieved by the first Russian tokamaks in 1968; and the heat generated by TFTR and JET in the 1990s which got close to break-even but didn't quite get there. Each time, the press has excitedly published accounts of the promise of fusion for solving the world's energy problems but then unexpected technical problems, lack of funding or simply the slow pace of research has meant that fusion has faded again from the public consciousness. Many have grown cynical that fusion will ever deliver on its promises. Remember the jibe: fusion is the energy of the future, and always will be.

But there are legitimate concerns that fusion will ever provide an economic source of energy – even if high gain is achieved – and those concerns are usually expressed by engineers. They argue that fusion scientists' fixation on developing a reactor that will simply produce more energy than it consumes ignores the very serious hurdles such a reactor would still have to overcome before it could compete with existing sources of energy.

In 1994 the Electric Power Research Institute (EPRI) – the R&D wing of the US electric utility industry – asked a panel of industry R&D managers and senior executives to draw up a set of criteria that a fusion reactor would have to meet in order to be acceptable to the electricity industry. They came up with three. The first is economics: to compensate for the increased risk of adopting a new technology, a new fusion plant would need to have lower life-cycle costs than competing technologies at the time. The second criterion was public acceptance: it would need to be something the public wanted and had confidence in. Finally the industry panel would want fusion to have a simple regulatory approval process: if the nuclear regulator required a lengthy investigation of the design or required that the reactor be sited far from population centres or be encased in a containment building, fusion's prospects could be seriously damaged.

One of the first to question the viability of fusion power generation was Lawrence Lidsky, a professor of nuclear engineering at the Massachusetts Institute of Technology and an associate director of its Plasma Fusion Center. By 1983 Lidsky had been working in plasma physics and reactor technology for twenty years and had formed some serious concerns about fusion's future. Colleagues at the Plasma Fusion Center were reluctant to talk about it, so Lidsky wrote an article for MIT's magazine *Technology Review* entitled 'The Trouble with Fusion.' Lidsky argued that, because of the inescapable physics of deuterium-tritium fusion, any fusion power plant is going to be bigger, more complex and more expensive than a comparable nuclear fission reactor – and so would fail EPRI's economic and regulatory criteria – and that complexity would make it prone to small breakdowns – failing the public acceptance criterion.

Lidsky's first criticism was with the choice of the fuel itself. When the fusion pioneers of the 1940s and '50s realised how difficult it was going to be to get to temperatures high enough to cause fusion, they naturally sought the fuel that would react the most easily – a mixture of deuterium and tritium. Reacting any other combination of light nuclei, such as deuterium and deuterium or hydrogen and helium-3, was just not conceivable with the technology of the day. So D-T became the focus of fusion research and scientists chose to ignore the fact that the reaction produces copious quantities of high-energy neutrons that would be a major headache for any working power reactor. Fission reactors produce neutrons too but the ones in a fusion reactor have a higher energy and so penetrate the structure of the reactor itself where they can knock atoms in the steel out of position. Over years of operation this neutron bombardment makes the reactor radioactive and weakens it structurally, which limits its life and means that any maintenance or repair is difficult or impossible to carry out with human beings.

Key parts of the reactor could instead be made with other metals that are more resistant to neutrons, such as vanadium, but that would increase the cost. Fusion scientists have long been aware of this issue and have sought other more exotic neutron-resistant materials but such efforts have always played second fiddle to the drive towards break-even. In any event, testing the materials would require a very intense source of neutrons and today no such source exists. US researchers have proposed building fusion reactors that are optimised to produce lots of neutrons rather than energy and one of these could be used as a testbed for new materials. But with the country's fusion budget severely squeezed it was never a high priority.

Another option is to produce neutrons using a high-intensity particle beam in a purpose-built accelerator facility. The agreement between the European Union and Japan over where to build ITER, the so-called broader approach, provided money to start work on such a test rig, dubbed IFMIF (International Fusion Materials Irradiation Facility), but at the time of writing this project was still working on a design and testing technology – far from actually bombarding any materials with neutrons. So the effort to find suitable materials for a fusion reactor lags behind the rest of the fusion enterprise and is unlikely to provide any useful data for ITER, though it could for ITER's successors.

Lidsky also contended that a D-T fusion reactor would be unavoidably large and complex and therefore unacceptable to the electricity industry. For a start, the reactor must cope with a range of temperatures that drops over the distance of a few metres from roughly 150,000,000°C – hotter than anywhere else in our solar system – to -269°C, a few degrees from absolute zero temperature, which is the operating temperature of the superconducting magnets. Managing these temperature gradients and heat flows will be a major challenge. And, he argued, a power-producing D-T reactor cannot avoid being large and therefore expensive. History backs up his assertion: as tokamaks have developed they have got bigger and bigger. The plasma vessel of ITER is 19m across and this is just the start: outside it are several metres more including the first wall – the initial line of defence against heat and neutrons – the 'blanket' through which liquid lithium will pass so that it can be bred into tritium by the neutrons, a thermal shield to protect the super-cold magnets from the heat of the reactor and finally the magnets themselves. Altogether a huge structure, considerably larger than the core of a fission reactor and in power engineering, size = cost. Anyway, ITER is not designed to generate electricity; the demonstration power reactor that is proposed to come after it, known as DEMO, will by some estimates be 15% larger in linear dimensions.

If fusion research continued on its current course, Lidsky concluded in 1983, 'the costly fusion reactor is in danger of joining the ranks of other technical "triumphs" such as the zeppelin, the supersonic transport and the fission breeder reactor that turned out to be unwanted and unused.' His *Technology Review* article was followed by an adapted version in the *Washington Post* entitled 'Our Energy Ace in the Hole Is a Joker: Fusion Won't Fly.' These public criticisms caused a furore in fusion research and led to a war of letters between Lidsky and PPPL director Harold Furth lasting many months. Shortly after the articles were published, Lidsky was stripped of his associate director title at the Plasma Fusion Center and he became a pariah in the fusion community.

But Lidsky's message was not entirely negative. He acknowledged the attractions of limitless fuel and minimal radioactive waste from fusion, but in essence he thought that fusion had taken a wrong turn and needed to start again by focusing on a different reaction that produces no neutrons: the fusion of hydrogen and boron-11. This seems ideal, but boron has five times the positive charge of hydrogen, making fusion much harder to achieve. Although some schemes for fusing hydrogen and boron have been proposed – including using Sandia's Z Machine – none have yet been tested.

Although these concerns were expressed three decades ago, many of them hold true today. Better materials and techniques exist now, but the basic physics remains the same. Fusion enthusiasts concede that there are some major challenges ahead but it is not that they can't be solved; they just haven't been solved yet. Just because something is hard, it doesn't mean we shouldn't attempt it. But even many of the most ardent fusion enthusiasts concede that commercial fusion power is unlikely before 2050. That view is supported by another more recent report from EPRI which sought to find out if any fusion technologies existing now might be useful to the power industry in the near term. Published in 2012, it examined magnetic as well as inertial confinement fusion approaches and some of the alternative schemes but concluded that all were in an early stage of technical readiness and none would be ready for use in the next thirty years. It suggested that fusion research efforts ought to pay more attention to the problem of generating electricity instead of fixating on the scientific feasibility of producing excess energy.

So is fusion the energy dream that is destined never to be fulfilled? Fusion is often compared to fission, both having been born out of the postwar enthusiasm for all things technological. Fission proved its worth astonishingly fast: the splitting of heavy elements was discovered in 1938; the first atomic pile produced energy in 1942; and the first experimental electricity-producing plant started in 1951. Thirteen years from discovery to electricity. But the comparison with fission is misleading: it hardly takes any energy at all to cause a uranium-235 nucleus to split apart and release some of its store of energy; and as a fuel, solid uranium is easy to handle. Fusion, in contrast, requires temperatures ten times those in the core of the Sun and its fuel is unruly plasma. When fusion research began, scientists had little real knowledge of how plasmas behave and, although great strides have been taken, there is still a lot we don't know about plasma physics.

Fission had another advantage in its early years: it was born into the breakneck wartime effort to develop an atomic bomb. The early atomic reactors were essential to that effort because they produced plutonium, so money was poured into their development. After the war, the rapid research continued as military planners saw fission reactors as a perfect power source for submarines. In fact, the light-water reactor, which has come to dominate the nuclear power industry, is little more than an adapted submarine power plant and, with the benefit of hindsight, wasn't the best choice for land-based electricity generation. All of that development would probably have taken decades if carried out by civilian scientists in peacetime.

Fusion was also born in the military research labs of Britain, the United States and the Soviet Union and at first was kept classified for the same reason as fission: because it could be used to create plutonium. But as soon as military planners realised it would be no easier than using a fission reactor, they lost interest in controlled fusion. Since then, fusion research has struggled along, buffeted by cycles of feast and famine as government enthusiasm for alternative sources of energy has waxed and waned. It has never enjoyed clear enthusiastic support from government such as the Apollo programme received following President Kennedy's pledge to put a man on the Moon. This is one reason for fusion research's obsession with achieving break-even. A clear demonstration that fusion can generate an excess of energy will grab the headlines and generate a wave of excitement. Scientists and engineers will then be clamouring to join the ranks of fusion researchers and government funding will come pouring in. That's the hope, anyway.

Some technological dreams just do take time to come to fruition. Look at the history of aviation. The Wright brothers' flight in 1903 wasn't the start of the process; aviation pioneers had been struggling to get aloft for decades before that. The first twelve-second flight at Kitty Hawk was just the demonstration of feasibility – analogous to break-even in fusion, perhaps? Orville and Wilbur had no idea how their invention would develop next. They couldn't possibly have imagined the huge jet airliners of today, let alone the spacefaring pleasure craft of Virgin Galactic. But those developments didn't happen instantly: it took more than four decades, and the accelerated development of two world wars, before jet engines and pressurised cabins became the norm. Fusion is still at the wooden struts, wire and canvas stage of development. Future fusion power plants may look nothing like a puffed up version of ITER.

The cost and time that it will take to make fusion work has to be balanced against the enormous benefits it would bring. Assuming that all the engineering hurdles described by the likes of Lidsky can be overcome, what would a world powered by fusion be like? The current partners in ITER represent more than half the world's population, so the technological know-how to build fusion reactors will be widespread – there will be no monopoly. Nor will any nation have a stranglehold over fuel supplies. Every country has access to water. There will be no more mining for coal or digging up tar sands; no more oil rigs at sea or in fragile habitats on land; no more pipelines scything across wildernesses; and no more oil tankers or oil spills. The geopolitics of energy – with all the accompanying corruption, coups and wars of access – will disappear. Countries with booming economic growth, such as China, India, South Africa and Brazil, will no longer have to resort to helter-skelter building of coal-fired and nuclear power stations. It is highly unlikely that fusion power will be 'too cheap to meter,' as US Atomic Energy Commission chief Lewis Strauss claimed in 1954, but it won't damage the climate, it won't pollute and it won't run out. How can we not try?

Getting there will not be easy and it won't happen unless society at large and its governments, as well as fusion scientists, want it. Lev Artsimovich, the Soviet fusion pioneer who led their effort for more than twenty years, was once asked when fusion energy would be available. He replied: 'Fusion will be ready when society needs it.'

# Further Reading

*The Scientific Origins of Controlled Fusion Technology*, by John Hendry, *Annals of Science*, vol. 44 (1987), pp. 143-168

An excellent summary of the 'prehistory' of fusion research and early work in the United Kingdom up to 1950.

*Fusion Research in the UK 1945-1960*, by J. Hendry and J. D. Lawson (AEA Technology, 1993)

A definitive history of fusion research's 'heroic age' in the United Kingdom.

*Project Sherwood: The U.S. Program in Controlled Fusion*, by Amasa S. Bishop (Addison-Wesley, 1958)

The story of America's early years, by someone who was there.

*Fusion: Science, Politics, and the Invention of a New Energy Source*, by Joan Lisa Bromberg (MIT Press, 1982)

The official history of the US programme from its beginnings to 1980.

*Fusion: The Search for Endless Energy*, by Robin Herman (Cambridge University Press, 1990)

A popular account of the international fusion effort up to the late 1980s.

*The Science of JET*, by John Wesson (JET Joint Undertaking, 2000)

A brief technical history of the Joint European Torus.

*Nuclear Fusion: Half a Century of Magnetic Confinement Fusion Research*, by C. M. Braams and P. E. Stott (Taylor & Francis, 2002)

A thorough history of magnetic fusion packed with technical details.

*Lasers Across the Cherry Orchards*, by Michael Forrest (Michael Forrest, 2011)

A personal memoir of the Culham scientists' expedition to Moscow to take the temperature of the T-3 tokamak.

*Inertial Confinement Nuclear Fusion: A Historical Approach by Its Pioneers*, by Guillermo Velarde and Natividad Carpintero-Santamaria (eds) (Foxwell & Davies, 2007)

A collection of personal accounts of the development of inertial confinement fusion.

*Search for the Ultimate Energy Source: A History of the U.S. Fusion Energy Program*, by Stephen O. Dean (Springer, 2013)

A thorough account of the US effort by one of its main protagonists.

# Acknowledgments

Firstly I would like to thank my colleagues at *Science* magazine who have aided and abetted my attempts to chronicle fusion's progress, including Colin Norman, Robert Coontz, John Travis, Richard Stone, Eliot Marshall, Jeffrey Mervis, Adrian Cho, Robert F. Service, Dennis Normile, and Andrey Allakhverdov.

This book would be nothing without the scores of scientists and science administrators who have given their time and effort to talk to me and explain the intricacies of their work. For that I would like to thank Roberto Andreani, Robert Aymar, Michael Bell, Stephen Bodner, Harald Bolt, Duarte Borba, Richard J. Buttery, David Campbell, Valery Chuyanov, Tom Cochran, John Collier, Bruno Coppi, Glenn Counsell, Michael Cuneo, Ron Davidson, Anne Davies, Stephen O. Dean, Arnaud Devred, Mike Dunne, Jacques Ebrardt, Chris Edwards, Umberto Finzi, Eric Fredrickson, Richard Garwin, David Gates, Alan Gibson, Siegfried Glenzer, Robert Goldston, Martin Greenwald, Greg Hammett, David Hammer, Norbert Holtkamp, Lorne Horton, Kaname Ikeda, Kimihiro Ioki, Jean Jacquinot, Gunter Janeschitz, Raymond Jeanloz, Bob Kaita, Marylia Kelley, Thomas Klinger, Russell Kulsrud, Joe Kwan, John Lindl, Steven Lisgo, Christopher Llewellyn-Smith, Dick Majeski, Guy Matthews, Keith Matzen, Robert McCrory, Dale Meade, Sergei Mirnov, Neil Mitchell, Achilleas Mitsos, Edward Moses, Osamu Motojima, Dennis Mueller, Vladimir Semenovich Mukhovatov, Steve Obenschain, Chris Paine, Jerome Pamela, Richard Pitts, Stewart Prager, Sergei Putvinski, Rezwan Razani, Ksenia Aleksandrovna Razumova, Paul-Henri Rebut, Michael Roberts, Francesco Romanelli, Steven Sabbagh, Ned Sauthoff, Roy Schwitters, John Sethian, Yasuo Shimomura, Jim Strachan, Vyacheslav Sergeevich Strelkov, Edmund Synakowski, Bryan Taylor, Paul Vandenplas, Evgeniy Velikhov, Michael Watkins, Peide Weng, Randy Wilson, Glen A. Wurden, Ken Young, Michael Zarnstorff and Hartmut Zohm. If I've forgotten anyone, please accept my apologies. Your contribution was no less valuable.

I'm also eternally grateful to the unsung heroes of science writing: lab and university press officers. My thanks go to Aris Apollonatos, Neil Calder, Chris Carpenter, Michel Claessens, Mark Constance, Sabina Griffith, Jennifer Hay, Bonnie Hébert, Judith Hollands, Nick Holloway, Kitta MacPherson, Isabella Milch, John Parris, Paul Preuss, Lynda Seaver, Jeff Sherwood, Bill Spears, Eleanor Starkman, Chris Warwick, Patti Wieser, and Mark Woollard.

Special thanks go to Steven Cowley, director of the Culham Centre for Fusion Energy, for reading the manuscript and helping me avoid some terrible mistakes, and to my agent Peter Tallack of the Science Factory for nursing this project through to publication.

I'm indebted to my editors, Jon Jackson at Duckworth and Dan Crissman at Overlook, for polishing up the manuscript so nicely, and to everyone there, including Tracy Carns, Michael Goldsmith, Peter Mayer and Jamie-Lee Nardone, for turning my words into such a fabulous book.

I'd also like to thank Adam, Giles, Jeremy, Jo, June, Marek, Megan, Natalie, Philip and Tig of the Ufford House Book Group for their six-weekly doses of encouragement.

And finally, love and thanks to Bernadette Lawrence, Sam Lawrence-Clery and Ellen Lawrence-Clery for your constant support and encouragement and for being the most exceptional family.

# INDEX

# INDEX

# INDEX

# INDEX